Reviews of Environmental Contamination and Toxicology

VOLUME 235

For further volumes:
http://www.springer.com/series/398

Reviews of Environmental Contamination and Toxicology

Editor
David M. Whitacre

Editorial Board
Maria Fernanda, Cavieres, Valparaiso, Chile • Charles P. Gerba, Tucson, Arizona, USA
John Giesy, Saskatoon, Saskatchewan, Canada • O. Hutzinger, Bayreuth, Germany
James B. Knaak, Getzville, New York, USA
James T. Stevens, Winston-Salem, North Carolina, USA
Ronald S. Tjeerdema, Davis, California, USA • Pim de Voogt, Amsterdam, The Netherlands
George W. Ware, Tucson, Arizona, USA

Founding Editor
Francis A. Gunther

VOLUME 235

Coordinating Board of Editors

Dr. David M. Whitacre, *Editor*
Reviews of Environmental Contamination and Toxicology

5115 Bunch Road
Summerfield, North Carolina 27358, USA
(336) 634-2131 (PHONE and FAX)
E-mail: dmwhitacre@triad.rr.com

Dr. Erin R. Bennett, *Editor*
Bulletin of Environmental Contamination and Toxicology
Great Lakes Institute for Environmental Research

University of Windsor
Windsor, Ontario, Canada
E-mail: ebennett@uwindsor.ca

Peter S. Ross, *Editor*
Archives of Environmental Contamination and Toxicology

7719 12th Street
Paron, Arkansas 72122, USA
(501) 821-1147; FAX (501) 821-1146
E-mail: AECT_editor@earthlink.net

ISSN 0179-5953   ISSN 2197-6554 (electronic)
ISBN 978-3-319-10860-5   ISBN 978-3-319-10861-2 (eBook)
DOI 10.1007/978-3-319-10861-2
Springer Cham Heidelberg New York Dordrecht London

© Springer International Publishing Switzerland 2015
This work is subject to copyright. All rights are reserved by the Publisher, whether the whole or part of the material is concerned, specifically the rights of translation, reprinting, reuse of illustrations, recitation, broadcasting, reproduction on microfilms or in any other physical way, and transmission or information storage and retrieval, electronic adaptation, computer software, or by similar or dissimilar methodology now known or hereafter developed. Exempted from this legal reservation are brief excerpts in connection with reviews or scholarly analysis or material supplied specifically for the purpose of being entered and executed on a computer system, for exclusive use by the purchaser of the work. Duplication of this publication or parts thereof is permitted only under the provisions of the Copyright Law of the Publisher's location, in its current version, and permission for use must always be obtained from Springer. Permissions for use may be obtained through RightsLink at the Copyright Clearance Center. Violations are liable to prosecution under the respective Copyright Law.
The use of general descriptive names, registered names, trademarks, service marks, etc. in this publication does not imply, even in the absence of a specific statement, that such names are exempt from the relevant protective laws and regulations and therefore free for general use.
While the advice and information in this book are believed to be true and accurate at the date of publication, neither the authors nor the editors nor the publisher can accept any legal responsibility for any errors or omissions that may be made. The publisher makes no warranty, express or implied, with respect to the material contained herein.

Printed on acid-free paper

Springer is part of Springer Science+Business Media (www.springer.com)

Foreword

International concern in scientific, industrial, and governmental communities over traces of xenobiotics in foods and in both abiotic and biotic environments has justified the present triumvirate of specialized publications in this field: comprehensive reviews, rapidly published research papers and progress reports, and archival documentations. These three international publications are integrated and scheduled to provide the coherency essential for nonduplicative and current progress in a field as dynamic and complex as environmental contamination and toxicology. This series is reserved exclusively for the diversified literature on "toxic" chemicals in our food, our feeds, our homes, recreational and working surroundings, our domestic animals, our wildlife, and ourselves. Tremendous efforts worldwide have been mobilized to evaluate the nature, presence, magnitude, fate, and toxicology of the chemicals loosed upon the Earth. Among the sequelae of this broad new emphasis is an undeniable need for an articulated set of authoritative publications, where one can find the latest important world literature produced by these emerging areas of science together with documentation of pertinent ancillary legislation.

Research directors and legislative or administrative advisers do not have the time to scan the escalating number of technical publications that may contain articles important to current responsibility. Rather, these individuals need the background provided by detailed reviews and the assurance that the latest information is made available to them, all with minimal literature searching. Similarly, the scientist assigned or attracted to a new problem is required to glean all literature pertinent to the task, to publish new developments or important new experimental details quickly, to inform others of findings that might alter their own efforts, and eventually to publish all his/her supporting data and conclusions for archival purposes.

In the fields of environmental contamination and toxicology, the sum of these concerns and responsibilities is decisively addressed by the uniform, encompassing, and timely publication format of the Springer triumvirate:

Reviews of Environmental Contamination and Toxicology [Vol. 1 through 97 (1962–1986) as Residue Reviews] for detailed review articles concerned with any aspects of chemical contaminants, including pesticides, in the total environment with toxicological considerations and consequences.

Bulletin of Environmental Contamination and Toxicology (Vol. 1 in 1966) for rapid publication of short reports of significant advances and discoveries in the fields of air, soil, water, and food contamination and pollution as well as methodology and other disciplines concerned with the introduction, presence, and effects of toxicants in the total environment.

Archives of Environmental Contamination and Toxicology (Vol. 1 in 1973) for important complete articles emphasizing and describing original experimental or theoretical research work pertaining to the scientific aspects of chemical contaminants in the environment.

Manuscripts for Reviews and the Archives are in identical formats and are peer reviewed by scientists in the field for adequacy and value; manuscripts for the Bulletin are also reviewed, but are published by photo-offset from camera-ready copy to provide the latest results with minimum delay. The individual editors of these three publications comprise the joint Coordinating Board of Editors with referral within the board of manuscripts submitted to one publication but deemed by major emphasis or length more suitable for one of the others.

Coordinating Board of Editors

Preface

The role of Reviews is to publish detailed scientific review articles on all aspects of environmental contamination and associated toxicological consequences. Such articles facilitate the often complex task of accessing and interpreting cogent scientific data within the confines of one or more closely related research fields.

In the nearly 50 years since *Reviews of Environmental Contamination and Toxicology* (formerly *Residue Reviews*) was first published, the number, scope, and complexity of environmental pollution incidents have grown unabated. During this entire period, the emphasis has been on publishing articles that address the presence and toxicity of environmental contaminants. New research is published each year on a myriad of environmental pollution issues facing people worldwide. This fact, and the routine discovery and reporting of new environmental contamination cases, creates an increasingly important function for *Reviews*.

The staggering volume of scientific literature demands remedy by which data can be synthesized and made available to readers in an abridged form. *Reviews* addresses this need and provides detailed reviews worldwide to key scientists and science or policy administrators, whether employed by government, universities, or the private sector.

There is a panoply of environmental issues and concerns on which many scientists have focused their research in past years. The scope of this list is quite broad, encompassing environmental events globally that affect marine and terrestrial ecosystems; biotic and abiotic environments; impacts on plants, humans, and wildlife; and pollutants, both chemical and radioactive; as well as the ravages of environmental disease in virtually all environmental media (soil, water, air). New or enhanced safety and environmental concerns have emerged in the last decade to be added to incidents covered by the media, studied by scientists, and addressed by governmental and private institutions. Among these are events so striking that they are creating a paradigm shift. Two in particular are at the center of everincreasing media as well as scientific attention: bioterrorism and global warming. Unfortunately, these very worrisome issues are now superimposed on the already extensive list of ongoing environmental challenges.

The ultimate role of publishing scientific research is to enhance understanding of the environment in ways that allow the public to be better informed. The term "informed public" as used by Thomas Jefferson in the age of enlightenment conveyed the thought of soundness and good judgment. In the modern sense, being "well informed" has the narrower meaning of having access to sufficient information. Because the public still gets most of its information on science and technology from TV news and reports, the role for scientists as interpreters and brokers of scientific information to the public will grow rather than diminish. Environmentalism is the newest global political force, resulting in the emergence of multinational consortia to control pollution and the evolution of the environmental ethic. Will the new politics of the twenty-first century involve a consortium of technologists and environmentalists, or a progressive confrontation? These matters are of genuine concern to governmental agencies and legislative bodies around the world.

For those who make the decisions about how our planet is managed, there is an ongoing need for continual surveillance and intelligent controls to avoid endangering the environment, public health, and wildlife. Ensuring safety-in-use of the many chemicals involved in our highly industrialized culture is a dynamic challenge, for the old, established materials are continually being displaced by newly developed molecules more acceptable to federal and state regulatory agencies, public health officials, and environmentalists.

Reviews publishes synoptic articles designed to treat the presence, fate, and, if possible, the safety of xenobiotics in any segment of the environment. These reviews can be either general or specific, but properly lie in the domains of analytical chemistry and its methodology, biochemistry, human and animal medicine, legislation, pharmacology, physiology, toxicology, and regulation. Certain affairs in food technology concerned specifically with pesticide and other food-additive problems may also be appropriate.

Because manuscripts are published in the order in which they are received in final form, it may seem that some important aspects have been neglected at times. However, these apparent omissions are recognized, and pertinent manuscripts are likely in preparation or planned. The field is so very large and the interests in it are so varied that the editor and the editorial board earnestly solicit authors and suggestions of underrepresented topics to make this international book series yet more useful and worthwhile.

Justification for the preparation of any review for this book series is that it deals with some aspect of the many real problems arising from the presence of foreign chemicals in our surroundings. Thus, manuscripts may encompass case studies from any country. Food additives, including pesticides, or their metabolites that may persist into human food and animal feeds are within this scope. Additionally, chemical contamination in any manner of air, water, soil, or plant or animal life is within these objectives and their purview.

Manuscripts are often contributed by invitation. However, nominations for new topics or topics in areas that are rapidly advancing are welcome. Preliminary communication with the editor is recommended before volunteered review manuscripts are submitted.

Summerfield, NC, USA

David M. Whitacre

Contents

Factors Influencing the Toxicity, Detoxification and Biotransformation of Paralytic Shellfish Toxins 1
Kar Soon Tan and Julian Ransangan

Sources, Fluxes, and Biogeochemical Cycling of Silver in the Oceans 27
Céline Gallon and A. Russell Flegal

DDT, Chlordane, Toxaphene and PCB Residues in Newport Bay and Watershed: Assessment of Hazard to Wildlife and Human Health 49
James L. Byard, Susan C. Paulsen, Ronald S. Tjeerdema, and Deborah Chiavelli

Index 169

Factors Influencing the Toxicity, Detoxification and Biotransformation of Paralytic Shellfish Toxins

Kar Soon Tan and Julian Ransangan

Contents

1 Introduction	1
2 Paralytic Shellfish Toxins	2
3 Factors that Influence HABs and their Toxicity	6
3.1 Eutrophication	6
3.2 Environmental Parameters	7
4 Accumulation and Depuration of PST in Bivalves	9
5 Analysis Methods for Paralytic Shellfish Toxins	10
5.1 Bioassay Methods	12
5.2 Chemical Assay Methods	13
5.3 Immunosorbant Assays	14
6 Biotransformation of PSTs	15
7 Conclusions	16
8 Summary	17
References	18

1 Introduction

The incidence of red tide events globally has escalated in marine coastal environments over the last several decades (Cordier et al. 2000). The term red tide is used to describe a phenomenon in which a water body exhibits red coloration from the presence of high algal cell density. Red tide events are often harmful to both human and aquatic organisms. However, the term may be confusing, because red tide refers not

---

K.S. Tan • J. Ransangan (✉)
Microbiology and Fish Disease Laboratory, Borneo Marine Research Institute,
University Malaysia Sabah, Jalan UMS, 88400 Kota Kinabalu, Sabah, Malaysia
e-mail: liandra@ums.edu.my

only to the high density of microscopic algal cells that colorize water, but also includes blooms of highly toxic cells that can cause problems even at low cell densities, i.e., a few hundred cells L^{-1}. Therefore, the term Harmful Algal Blooms (HABs) has been introduced to describe blooms of both toxic and non-toxic algae that potentially have negative effects on humans and the environment (Anderson 2009). The reported global incidence of paralytic shellfish poisoning (PSP) that has been associated with HABs has been increasing annually (Anderson 1989). However, it is still unclear whether the increase results from elevated public awareness and reporting of HABs, or from an increase in anthropogenic factors, like increasing marine pollution incidents.

Under normal conditions, HAB cells in aquatic environments exist at a negligible concentrations, and therefore, are harmless. However, certain conditions trigger harmful algal cell blooms, resulting in the secretion of high concentrations of potent toxins, e.g., saxitoxin (STX) (Deeds et al. 2008). All aquatic organisms are vulnerable to saxitoxin except bivalves (e.g., mussels, clams, scallops, oysters, and butter clams). This is because bivalves are filter feeders that consume plankton, which are then transferred from gills to digestive organs wherein the toxins are concentrated (Soon and Ransangan 2014; Bricelj and Shumway 1998).

Prolonged exposure to PSTs may produce mutations that increase the resistance of shellfish to the toxin (Brivelj et al. 2005). When this occurs, the mutated shellfish are capable of accumulating toxins to a fatal concentration. PSP is caused by ingestion of shellfish that are contaminated with one or more PST chemicals. Clinical symptoms of PST intoxication include numbness of the lips, tongue and fingertips, within minutes after consumption, followed by numbness of neck, legs and arms, combined with a feeling of lightness, dizziness and headache (Acres and Gray 1978). Death by respiratory failure can occur within 1–24 h post-ingestion (Rodrigue et al. 1990). However, for those who survive intoxication, no lasting effect remains after 24 h. Several countries have established a regulatory action limit of 80 µg STXeg/100 g of shellfish flesh, which is equivalent to 0.2 g STXeq MU^{-1} for STX in consumed food (AOAC 1995).

2 Paralytic Shellfish Toxins

PSTs are typically comprised of STX and STX-related compounds, and these are the toxins that cause fatal PSP. The PSTs are water soluble neurotoxins that are responsible for causing 50,000–500,000 human intoxications annually, and are estimated to cause a 1.5% mortality rate (Wang et al. 2008). PSTs cause a highly reversible and specific sodium channel blockage of ion transport when present at an equimolar ratio. The blockage is proposed to result from the binding of a positively charged 7,8,9-guanidinium group of STX to the negatively charged carboxyl groups at site 1 of the sodium channel (Cestele and Catterall 2000). This binding prevents conductance of neuronal signals and causes muscular paralysis. In severe cases, death may result from respiratory failure (Baden and Trainer 1993). To date, there is no antidote for PSP. However, the symptomatic treatments for PSP that are commonly used include the following: fluid therapy, blood pH monitoring, and artificial respiration with benzedrine administration, and gastric lavage with dilute bicarbonate solution (Kao 1993). Su et al. (2004) discovered that PSTs also bind to the human potassium channel. However, the toxins modify the potassium channel, rather than blocking it as occurs in sodium channels. To date, four or more PST molecules are known to bind to extracellular sites of the potassium channel and to produce strong transmembrane depolarization (Wang et al. 2003). Depolarization causes the channel to open and to reduce potassium conductance. PSTs also act on voltage-gated calcium channels, by incompletely blocking them at extracellular sites (Su et al. 2004).

In general, saxitoxin and its derivatives are grouped into three categories; carbamate toxins, decarbamoyl toxins, and N-sulfocarbamoyl toxins. The most potent carbamate toxins include saxitoxin (STX), neosaxitoxin (neoSTX) and gonyautoxins (GTX1-4). The decarbamoyl analogues (viz., dcSTX, dcNEO, dcGTX1-4) are reported to have intermediate toxicity, whereas the least toxic derivatives are the N-sulfocarbamoyl toxins (B1 (GTX5), B2 (GTX6) and C1-C4) (Llewellyn 2006; Lassus et al. 2000). The basic structure of the PSTs comprises a trialkyl tetrahydropurine skeleton, with positions 2 and 8 of the purine ring containing NH_2 groups (Schantz et al. 1975). The 1,2,3-guanidine moiety carries a positive charge, whereas the 7,8,9-guanidine group is partially deprotonated. In Table 1, we present the molecular structures of the PSTs, and provide information on their relative toxicity. There are four functional groups (R1-4) clustered around the PST ring (Table 1). These functional groups determine the toxic potency of STX.

PSTs are found in both marine and freshwater environments and are produced by organisms inhabiting two taxonomic kingdoms: the prokaryotic cyanobacteria and eukaryotic dinoflagellates. In marine habitats, the PSTs are produced by bloom-forming dinoflagellates such as: *Gymnodinium catenatum* (Oshima et al. 1993), *Pyrodinium bahamense* Var. *compressum* (Usup et al. 1994, 2002) and *Alexandrium* spp. (Lim et al. 2005; Lefebvre et al. 2008). In Table 2, we present a grid that shows what PSTs are produced by different dinoflagellate organisms. As shown in Table 1, the different PSTs possess different levels of toxicity. *Gymnodinium catenatum* is widely distributed in the Gulfs of California and Mexico, Argentina, Japan, the Philippines, Palau, Tasmania, the Mediterranean, the Atlantic coast of Morocco, Portugal, and Spain (Hallegraeff 1993). *Pyrodinium bahamense* Var. *compressum* is largely responsible for PSP outbreaks in tropical waters of the Indo-West Pacific (Sabah, Brunei, Philippines and Papua-New Guinea) (Usup et al. 1994) and East Pacific (off the coast of Guatemala and Mexico) (Hallegraeff 1993). *Alexandrium* spp. is widely distributed in both temperate and tropical waters of the Pacific coast, from Alaska to southern California (Taylor 1984), Argentine Sea (Carreto et al. 1986), Magallanes Strait (Benavides et al. 1995), Arabian Sea (Sharma et al. 2011), peninsular Malaysia (Lim et al. 2001), Northeastern Canada (Parkhill and Cembella 1999), Gulf of Thailand (Kodama 1990), Japan (Ogata et al. 1989), southern Australia, and New Zealand (Bricelj and Shumway 1998).

Table 1 Molecular structure and toxicity of paralytic shellfish toxins (PSTs) (*Source*: Wiese et al. 2010; Donovan et al. 2009; Smith et al. 2001; Oshima 1995)

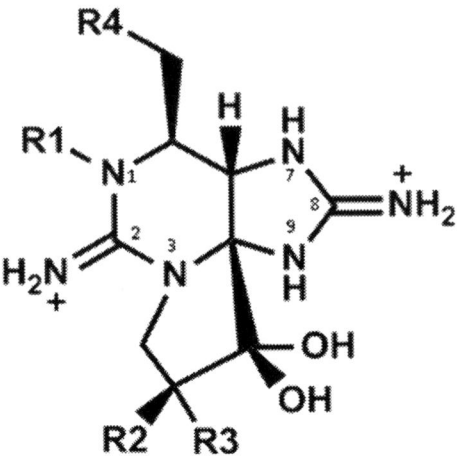

Toxins	Toxicity (MU µmol^{-1})	Functional groups			
		R1	R2	R3	R4
Carbamate					
STX	2,483	H	H	H	CONH$_2$
NEO	2,295	OH	H	H	CONH$_2$
GTX 1	2,468	OH	H	OSO$_3$-	CONH$_2$
GTX 2	892	H	H	OSO$_3$-	CONH$_2$
GTX 3	1,584	H	OSO$_3$-	H	CONH$_2$
GTX 4	1,803	OH	OSO$_3$-	H	CONH$_2$
Decarbamoyl					
dcSTX	1,274	H	H	H	OH
dcneoSTX	734	OH	H	H	OH
dcGTX 1	Unknown	OH	H	OSO$_3$-	OH
dcGTX 2	367	H	H	OSO$_3$-	H
dcGTX 3	925	H	OSO$_3$-	H	H
dcGTX 4	Unknown	OH	OSO$_3$-	H	OH
N-Sulfocarbamoyl					
GTX 5 (B1)	160	H	H	H	CONHSO$_3$-
GTX 6 (B2)	158	OH	H	H	CONHSO$_3$-
C 1	15	H	H	OSO$_3$-	CONHSO$_3$-
C2	329	OH	OSO$_3$-	H	CONHSO$_3$-
C3	33	OH	H	OSO$_3$-	CONHSO$_3$-
C4	143	OH	OSO$_3$-	H	CONHSO$_3$-

Note: MU = Mouse Units

Table 2 Harmful algal bloom species and the documented PSTs they produce

Paralytic shellfish toxins

HABs Species	Carbamate						N-Sulfocarbamoyl						Decarbamoyl					
	STX	neoSTX	GTX1	GTX2	GTX3	GTX4	GTX5	GTX6	C1	C2	C3	C4	dsSTX	dcneoSTX	dcGTX1	dcGTX2	dcGTX3	dcGTX4
A. andersoni	√																	
A. catenella	√	√	√	√	√	√	√	√	√	√		√	√			√	√	
A. circinalis	√			√	√		√		√	√			√			√	√	
A. fundyense	√	√	√	√	√	√	√	√	√	√						√	√	
A. minutum			√	√	√	√												
A. tamarense	√	√	√	√	√	√	√	√	√									
A. taylor	√	√		√	√	√		√		√			√					
A. ostenfeldii				√	√			√	√	√								
A. peruvianum	√		√	√		√	√	√	√	√		√	√					
Gymnodinium catenatum	√		√		√	√	√		√	√	√	√			√	√	√	√
Pyrodinium bahamense	√	√					√	√					√					

Source: *A. andersoni* (Ciminiello et al. 2000), *A. catenella* (Krock et al. 2007; Sebastian et al. 2005; Siu et al. 1997), *A. circinalis* (Teste et al. 2002; Velzeboer et al. 2000; Negri and Jones 1995), *A. fundyense* (Jaime et al. 2007; Poulton et al. 2005; Anderson et al. 1990), *A. tamarense* (Dell'Aversano et al. 2008; Ichimi et al. 2002; Parkhill and Cembella 1999), *A. minutum* (Pitcher et al. 2007; Hwang et al. 2003; Franco et al. 1994), *A. taylor* and *A. peruvianum* from (Lim et al. 2005), *A. ostenfeldii* (Hansen et al. 1992), *Gymnodinium catenatum* (Negri et al. 2007; Garate-Lizarraga et al. 2005; Holmes et al. 2002) and *Pyrodinium bahamense* from (Usup et al. 1994)

3 Factors that Influence HABs and their Toxicity

3.1 Eutrophication

Eutrophication is defined as the introduction of excessive nutrients from anthropogenic activities into the aquatic environment that increases algal biomass. Enrichment by phosphorus (P) and nitrogen (N) in a habitat is the most important contributor to eutrophication. In freshwater, phytoplankton production is strongly correlated with total phosphorus inputs, whereas nitrogen most influences primary producers in marine waters (Anderson et al. 2002). Over the last decade, human activities have resulted in a significant increase of nutrient releases to coastal ecosystems. In general, the concentrations of P and N in the ocean are estimated to have increased by three and fourfold, respectively, compared to levels recorded during pre-industrial and pre-agricultural eras (Smil 2001). HAB species, like other photosynthetic organisms, require nutrients for growth. The nutrients' source of HAB is not limited to dissolved inorganic nutrients (nitrate and phosphate), but also include dissolved inorganic and organic compounds, and particulate nutrients in the form of detritus (Anderson et al. 2002). Nutrient enrichment enhances the impact of toxic HABs in two ways. First, it increases the abundance of all phytoplankton, without altering the relative fraction of the total phytoplankton biomass. Second, the harmful effects of HAB increase as the abundance of toxic HAB species increases (Poulton et al. 2005). Nutrient enrichment may favor one or a few species to become dominant, and thereby to overwhelm other phytoplankton (Anderson et al. 2002).

Many photosynthetic bacteria and flagellate species have been documented to demonstrate physiological adaptation, which allows them to acquire both C and N via particle ingestion, or to consume dissolved organic compounds (Al-Azad et al. 2013; Graneli and Carlsson 1998). Moreover, *Alexandrium* sp. is physiologically adapted to take excess P up that is required for immediate growth at an early growth stage (Lim et al. 2010). Harmful dinoflagellates tend to occur in water that has seasonally high phosphate and nitrate level, high amounts of dissolved organic carbon (DOC), and other forms of organic nutrients (Glasgow et al. 2001; Glibert et al. 2001). Bloom development coincides with at least a threefold reduction in DOC and dissolved organic nitrogen (DON), which indicates that HAB cells have the ability to acquire nutrients via extracellular oxidation or hydrolysis (Lewitus et al. 2001). Nevertheless, brown tide species such as *Aureoumbra lagunensis* have been shown to be incapable of nitrate uptake, and thus must use a reduced form of N (Deyoe and Sutrle 1994). In fact, food-vacuole-containing prey fragments were found in *Dinophysis norvegica* and *D. acuminate* (Jacobson and Anderson 1996), *Heterosigma carterae*, *A. tamarense* (Nygaard and Tobiesen 1998), and *Gyrodinium galatheanum* (Li et al. 2000, 2001), which confirm these species to be mixotrophic (i.e., they have the ability to ingest particulate food). In addition, the grazing rate of flagellate phytoplankton has been shown to increase when the dissolved phosphate concentration is low (<0.05 µM). Several algal flagellate species (viz., *Alexandrium tamarense*, *Gyrodinium galatheanum*, *Chrysochromulina polylepis*, *Chrysochromulina Arizona*, *Prymnesium parvum*, *Ochromonas minima*, *Pseudopedinella* sp., and *Heterosigma akashiwo*) were reported to graze on bacteria to acquire phosphate, when inorganic nutrients are limited (Nygaard and Tobiesen 1998).

HAB cells become more abundant when nutrients increase, although the resultant cellular toxicity may be either higher or lower than under non-eutrophic conditions. Proctor et al. (1975) demonstrated that the toxin concentration in HAB cells is inversely associated with the mean growth rate of cells. However, Kodama (1990) reported no correlation between growth rate and cellular toxicity. Generally, altering the nutrient ratio affects the growth and cellular toxicity of HAB cells. *Alexandrium* sp. grow optimally at N/P ratios of 0.18–0.38 (Lim et al. 2010). Whereas, other HAB species, such as *Prorocentrum micans*, *P. sigmoides*, and *P. triestinum*, grow optimally at an N/P ratio of 5–15 (Hodgkiss and Ho 1997). In addition, the nutrient ratio also influences the cell size of HABs. Increased N/P ratios produce increased cell size of *A. minutum* due to the arrest of cells in the G1 phase (i.e., they do not undergo cell division), while other non-P compounds continued to be synthesized (Vaulot et al. 1996).

The amount of toxin produced by HAB species is affected by the types and concentration of nutrients to which they are exposed. The dinoflagellate *D. acuminata* produced enhanced toxin levels during N and P deprivation; the enhancement was sixfold larger during N-deprivation (Johansson et al. 1996). In contrast, *Chrysochromulina polylepis* produced toxin that was sixfold more toxic under P enrichment than during N-limited conditions (Johansson and Graneli 1999). Moreover, saxitoxin production in *A. tamarense* was five to tenfold higher during P-deprivation, than when deprived of N (Lippemeier et al. 2003; Anderson et al. 1990). Enrichment in organic nutrients also influences cellular toxicity of HABs. For example, increasing the dissolved organic nutrients concentration (urea level up to 1 mM) stimulated toxin production in *K. brevis* by sixfold and caused *K. brevis* to switch from autotrophic to heterotrophic nutrition (Shimizu et al. 1993). Lim et al. (2010) observed a similar phenomenon with *A. minutum*, in which ammonium induced toxin production under a P-deprivation condition. However, the N/P ratio has no effect on the toxin composition of *A. minutum*.

3.2 Environmental Parameters

The nutrient supply does not necessarily correlate with the rate at which nutrients are assimilated by phytoplankton, because the nutrient uptake capability of organisms vary (Anderson et al. 2002). Diatoms physiologically adapt to better exploit a high concentration of nitrate (Lomas and Glibert 2000). Therefore, diatom growth is highly correlated with ambient nitrate concentration. How effectively nutrients are assimilated by phytoplankton and the resultant effects on organism growth depend heavily on several environmental factors, e.g., salinity, light, temperature, and water column stability.

There is evidence that environmental parameters play a major role in modulating the growth and cellular toxicity of HABs. In Sabah, blooms of *Pyrodium bahamense var. compressum* appeared to be associated with increased salinity (Anton et al. 2000). Moreover, the growth rate of *A. tamarense* is relatively low at low salinities (viz., 10–15 ppt), but increased with increasing salinity (Lim and Ogata 2005). Parkhill and Cembella (1999) demonstrated a positive correlation between cellular toxicity and salinity in *A. tamarense*, where the highest cellular toxicity occurred at 30 ppt. Toxin content in *A. minutum* (Grzebyk et al. 2003) and *A. tamarense* (Parkhill and Cembella 1999) decreased with increasing salinity (Lim and Ogata 2005). This resulted from the nutrients allocated for growth being reduced at higher salinity, which limited nutrient available for toxin production (Lim and Ogata 2005).

The growth rate of cultured *Alexandrium* sp. increased as light increased up to a saturation point of 150 μmol m^{-2} s^{-1} at 20 °C (Parkhill and Cembella 1999). However, the maximum growth rates could be reached at lower irradiances if temperatures were higher (Ogata et al. 1990). A Malaysian *A. tamiyavanichii* strain achieved optimum growth when cultured under a light intensity of 40 μmol photons m^{-2} s^{-1} and at an optimum temperature of 25 °C (Lim et al. 2006). In other HAB species, blooms of *Karenia* have been associated with warmer and constant water temperature conditions (Dahl and Tangen 1993). In one Tunisian lagoon, a *Gymnodinium aureolum* bloom was associated with decreasing day length (Romdhane et al. 1998). Some study results suggested that the toxin content in *Alexandrium* sp. (Hamasaki et al. 2001; Ogata et al. 1989) and *Gymnodinium catenatum* (Ogata et al. 1989) was inversely proportional to the growth rate; slow growth at low temperature and under low light conditions stimulated toxin synthesis. The results of other studies indicated that the toxin produced by *A. tamarense* and *A. minutum* was not influenced by growth of the organism, but rather by exogenous factors (Kodama 1990). Parkhill and Cembella (1999) reported that the cellular toxicity of the toxin produced by *A. tamarense* was at a maximum under intermediate light conditions of 60 μmol m^{-2} s^{-1} (Parkhill and Cembella 1999). However, toxin production is suppressed in *Pyrodinium bahamense var. compressum* (Usup et al. 1995) and *A. minitum* (Lim et al. 2006) when they are grown under low light and low temperature conditions, due to the inhibition of the enzyme Ribulose-1, 5-bisphosphate carboxylase/oxygenase (RUBISCO).

In general, the PST composition produced by *Alexandrium* sp. is not influenced by the brightness of light to which the organism is exposed (Parkhill and Cembella 1999). However, temperature is an important factor in altering the PST composition in *A. tamiyavanichii*, where the composition of GTX4 increased gradually as cells were grown at temperatures higher than the optimum needed for growth (Lim et al. 2006). Transformation of GTX3 and GTX2 into GTX4 by enzymatic oxidation is enhanced by increased temperatures until the optimum temperature of 30 °C is reached (Kodama 2000). Salinity is another factor that influences the composition of PST in HAB cells. Low salinity stimulated *Alexandrium minimum* to produce a higher concentration of GTX1, which is the major toxin produced during periods of contamination (Hwang and Lu 2000). PSTs are stable structurally under acidic condition, but are easily oxidized under alkaline conditions; indeed, alkaline treatment reduces overall toxicity (WHO 1984). However, saxitoxin has a greater effect at neutral pH than under acidic or basic conditions (Baden and Trainer 1993). Both the guanidinium and hydroxyl groups of the toxins are required for sodium channel recognition. Changing the pH value of water affects these groups and can decrease or eliminates the biological effects of many PST toxins.

4 Accumulation and Depuration of PST in Bivalves

Filter feeding bivalves are known to be the principal vectors for PSP. However, different bivalve species have different toxic accumulation rates and capacities. For examples, *M. edulis* and *M. californianus* are known to accumulate high toxin levels in a short time (about 2 weeks) and to produce toxicity that is 2–4 times higher than other co-occurring species (Larocque and Cembella 1991; Hurst and Gilfillan 1997). The PST accumulation that occurs among bivalve species is associated with their variable tolerance to the toxins. Interspecific variability in resistance may be related to the species-specific binding characteristics of the polypeptidic receptor sites in the sodium channel, or to production of STX-binding proteins (Daigo et al. 1988; Mahar et al. 1991). Bivalve species that are insensitive to PSTs readily feed on toxic cells, and therefore accumulate toxins to extremely high levels (Bricelj et al. 1990). In contrast, species sensitive to PST exhibit physiological and behavioral responses such as shell clapping in scallops and complete shell valve closure in mussels that enable them to reduce the feeding rate to minimize exposure to toxic cells (Gainey and Shumway 1988). Purple clam was documented to attain a maximum toxin level in ~20–24 days by feeding on toxic algae. Excess toxin in algae was excreted as pseudofaeces and the toxin did not enter the digestive gland once the purple clam accumulated its maximum capacity of toxin (Chou et al. 2005).

In sensitive bivalves, the bioaccumulation of PSTs is enhanced when the diet of the producing organism is comprised of both toxic and nontoxic phytoplankton. The presence of nontoxic cells stimulates the bivalves to open their valves, which, in turn, promotes feeding on toxic cells (Bricelj et al. 1996). The size and body mass of bivalves also influence toxin accumulation rates. The weight-specific toxicity of bivalves (μg STXeq 100 g^{-1}) has been shown to be inversely associated with body size, because smaller individuals have higher cell ingestion rates per unit biomass than do larger bivalves. The small (3–4 cm in shell length) *M. edulis* attained peak toxicities twice as high as those of larger (>6 cm) mussels (Bricelj and Shumway 1998). The distribution among organs of toxin accumulation varied. During the toxification phase, the digestive gland or viscera was first to come into contact with toxic cells. Therefore, the digestive system absorbs and accumulates the greatest proportion of the total body burden of the toxin. Most researchers have concluded that bivalve viscera and hepatopancreas accumulate the majority of total toxins (80–90%), whereas the locomotory tissue (adductor muscle, and the muscular foot) of bivalves typically contains <3% of the PSP toxin body burden, despite its large contribution to total body mass (Choi et al. 2003; Bricelj and Shumway 1998).

Toxicity levels vary among bivalve species from differences in both toxin retention and rates of depuration (Deeds et al. 2008). Most bivalve species (e.g., Eastern oyster (*Crassostrea virginica*), Northern quahogs (*Mercenaria mercenaria*) and green mussel (*Perna viridis*)) remove PSTs relatively quickly (within a few weeks) (Gacutan et al. 1989). Other bivalves species, such as Washington butterclams, sea scallops (*Placopecten magellanicus*), and Atlantic surfclams (*Spisula solidissima*) are well known to be slow (up to 2 years) eliminators of the toxins, because the PSTs strongly bind to their siphon tissues (Beitler and Liston 1990). Generally, fast detoxifiers are those species that remove or detoxify the toxins at rates between 6% and 17% per day. At such rates, safe levels are attained within 1–10 weeks (Table 3). In contrast, slow detoxifiers (those that exhibit toxin elimination rates of ≤1%/day) take several months or even years to attain safe levels (Bricelj and Shumway 1998). Detoxification in bivalves begins with a fast initial depuration phase, followed by a slow depuration phase (Chou et al. 2005; Xie et al. 2013; Choi et al. 2003). The initial fast depuration results from evacuation of unassimilated toxin and toxigenic cells, whereas the slower depuration phase represents loss of toxins that had been assimilated and incorporated into tissues (Silvert et al. 1998).

The PST level in each organ does not remain constant over time. During toxification, the proportion of toxin contained in bivalve viscera remains fairly constant, but decreases steadily during the detoxification process (Donovan et al. 2008; Kwong et al. 2006; Martin et al. 1990). The hepatopancreas and viscera exhibit the highest rate of toxin elimination, whereas the foot and adductor muscles display the slowest rate (Bricelj et al. 1991). The different clearance rates results from the PSTs having been redistributed from the hepatopancreas to other tissues during the detoxification process (Bricelj and Cembella 1995). The non-visceral tissues of purple clam begin to accumulate PST (i.e., GTX) on day 16–20 from the initial exposure as a result of translocation from the digestive gland to muscle tissue (Chou et al. 2005). The Purple Clam transfers toxins from the digestive gland to the foot and adductor muscles earlier than other clam species, because the half-life of the PSP toxin is shorter in this species (Chou et al. 2005; Chen and Chou 2001). In the surfclam, *S. solidissima*, the rank order of various tissue pools, in terms of detoxification rates, is as follows: viscera > gill > foot adductor muscle (Bricelj and Cembella 1995). The hepatopancreases, viscera, gills, foot and adductor muscles of green mussels were reported to accumulate 40%, 34%, 17% and 10% of PST, respectively, after 7 days of detoxification (Kwong et al. 2006). Accumulation rates of PST in different compartments also depend on the types of toxin involved (Kwong et al. 2006; Bricelj and Cembella 1995). For example, elimination of C2 toxin was shown to be efficient from all organs. However, elimination of GTX4 was relatively slow from gills and viscera, because the toxin was retained by those organs. Kwong et al. (2006) reported that the elimination of STX and neoSTX from foot and adductor muscles was delayed because the PSTs were redistributed from other organs.

5 Analysis Methods for Paralytic Shellfish Toxins

PST analysis is mainly achieved through bioassay (*in vivo* mouse bioassays), *in vitro* cell assays, *in vitro* functional assays (receptor binding assay), chemical assays (HPLC and LC-MS) and immunosorbant assays (immunoassays).

Table 3 Detoxification of PSTs by bivalve molluscs (adults unless indicated), as measured by the time required to attain a regulatory level (RL) of 80 g STXeq 100 g^{-1}

	Species	Time to RL (weeks)	Detoxification rate (%/day)	Location
Fast to moderate detoxifiers:				
Clams	*Tresus capax*	5.2<t<11.6	-	BC, Canada
	Mercenaria mercenaria	3.4–6.1	9.3–9.5	L
	Meretrix casta	4.4	-	Kumble Estuary, India
	Mya arenaria	1.0–4.0	9.8	Maine, USA
		4.7	7.7	St. Lawrence Estuary, Can.
Mussels	*Mytilus edulis*	0.6–9.6	10.7	Maine, USA
		5.1	13.8	Gasp Bay, Canada
	Mytilus californianus	2.8–9.0	6.6–7.4	Pacific N. America
	Choromytilus palliopunctatus	1.2<t<7.0	17.2	Oaxaca, Mexico
	Perna viridis	1.7	9.3	L
	Modiolus modiolus	0.9–9.2	7.0	Maine, USA
Oysters	*Crassostrea gigas*	0.6–2.0	-	Pacific N. America
	Crassostrea iridescens	1.8–3.8	8.9–18.1	Oaxaca, Mexico
	Crassostrea cucullata	6.9	5.5	Kumble Estuary, India
	Ostrea edulis	6.4	4.0	Maine, USA
	Pecten maximus (scallop)	6.4	7.4	L
Slow detoxifiers:				
Clams and cockles	*Saxidomus giganteus*	114	0.3–0.6	BC, Canada
	Saxidomus nuttalli	>73	0.9	CA, USA
	Spisula solidissima	96–100	0.8–1.9	ME, USA
		>69	0.4–1.6	Georges Bank, USA
	Cardium edule	11.2	3.3	Portugal
	Soletellina diphos	>51	1.2–3.6	Tungkang, Taiwan
Scallops	*Placopecten* Dig.+mantle+gill	>28	0.2–0.6	L
	magellanicus Dig.	>5.2	0.6	L
	Patinopecten yessoensis Dig.	>12	1.2–4.1	Funka Bay, Japan
		>17	3.8	Ofunato Bay, Japan
Mussel	*Aulacomya ater* (mussel) Dig.	17	1.7–1.8	Southern Chile, Peninsula

Values apply to whole tissues unless otherwise specified (*Source*: Bricelj and Shumway 1998)
Notes: RL=Regulatory level; F=field; L=laboratory; Dig.=digestive gland; juv.=juveniles; -= no data

5.1 Bioassay Methods

5.1.1 In Vivo Mouse Bioassays

Mouse bioassay (MBA) is the only assay that measures the biological response of whole animals, thus allowing some correlation with human toxic effects. This method was first applied by Sommer and Meyer in 1937. The MBA is an internationally recognized analysis method for PST, and has been standardized by the Association of Official Analytical Chemists to offer quick and accurate measurements (AOAC 1995). The advantages of this assay are that it is cheap, simple and accurate. However, it does not provide information on the specific toxin composition. Moreover, the use of live animals in this procedure is required and therefore is a growing challenge vis-a-vis animal ethics. The results of this mouse bioassay can be confounded by the presence of high salt content, the presence of metals and gender of the mouse. A high salt concentration suppresses the toxic effects by 20–50% (Schantz et al. 1958), while the presence of zinc results in overestimates of toxicity (Aune et al. 1998). Aune et al. (2008) found that the neurotoxin concentration required to kill a female mouse was sevenfold higher than that for a male mouse of the same weight.

5.1.2 In Vitro Cell Assays

The basic principle of *in vitro* cell assays is to lace two biologically active drugs (veratridine and ouabine), in combination, into an established neuroblastoma cell line culture, then add the toxin that is to be tested. Veratridine is a sodium channel activator that causes influx of sodium, leading to cell swelling and eventually cell lysis. Ouabine blocks the action of Na/K-ATPases. The presence of the toxin, e.g., tetrodotoxin, acts as a sodium channel blocker, preventing the influx of sodium ions and thus protecting the mouse neuroblastoma cells. Vital staining is used to determine cell viability, and the intensity of color is proportional to the amount of toxin present (Jellett et al. 1992). The commercially available and shippable kit (MIST: Maritie *In vitro* Shellfish Test) available for performing this analysis has a 3 week shelf life, and was developed by Jellet Biotek Ltd. (Dartmouth, Nova Scotia; Canada) from a neuroblastoma cell bioassay (Jellett et al. 1998). Three versions of the MIST kit are available, and they vary by having different aims, detection accuracy and requirements for detection equipment. The Mini-MIST™ kit can only be used for pre-screening for the presence or absence of saxitoxin. The semi-quantitative version is designed to detect whether toxins in a sample exceed the regulatory limit. The fully quantitative version of MIST has a sensitivity of 2 µg STXeq/100 g for shellfish tissue, which is 20-fold more sensitive than the standard mouse bioassay (Jellett et al. 1992). This method, however, does not provide information on the specific toxin composition and has a short shelf life of only about 2–3 weeks.

5.1.3 Receptor Binding Assay

The receptor binding assay was first introduced in 1984 (Davio and Fontelo 1984) and was later developed into a Microtiter plate format (Vieytes et al. 1993; Doucette et al. 1997). This assay is based on the interaction between the toxins and one of the pharmacological targets (viz., site 1) of the sodium channel. This assay is a competitive displacement assay, in which radiolabelled tritiated saxitoxin (^{3}H-STX) competes with unlabeled STX and its derivatives for a given number of receptor sites of the voltage-gated sodium channel in a rat membrane preparation. When binding equilibrium is established in this assay, the radiolabelled saxitoxin is quantified by liquid scintillation counter. The percentage of receptor-bound radiolabelled saxitoxin is directly proportional to the amount of unlabeled toxin present in a standard or in a sample. Receptor binding affinity occurs in the descending order of STX, GTX1/4, neoSTX, GTX2/3, dcSTX, GTX5, which is similar to the order of toxicity obtained via the MBA. Because the affinity of a toxin for a receptor is directly proportional to its potency, this method yields a response that is representative of the integrated potencies of all PSP toxins present, and can thus be correlated with human response. Using the Microtiter plate format, a detection limit of about 4 μg STXeq/mL can be achieved (Doucette et al. 1997). However, as in the case of *in vitro* cell assays, this method does not provide information on the specific toxin composition, and is limited by its requirement for a scintillation counter and radioisotopic forms of the toxins being assayed.

5.2 *Chemical Assay Methods*

5.2.1 High Performance Liquid Chromatography (HPLC) with Fluorescence

High performance liquid chromatography (HPLC) is a chemical assay method that relies on the separation of toxins by using ion-interaction chromatography and a post-column reactor. In this method, the column effluent is oxidized to produce readily detectable derivatives for analysis by absorption or fluorescence detectors. HPLC with fluorescence is the most commonly used method for PST analysis. Two types of chromatographic methods (pre and post column derivatization) are employed, and each involves treating (derivatizing) the columns in a way that optimizes toxin separation. Using both methods requires an entire set of STX analogs to be prepared as external reference (standards). However, standards prepared by different suppliers produce variability of up to 20% in STX concentration response (Quilliam and Janecek 1993). Therefore, to achieve quality results, selecting a reliable supplier of standards is critical and the standard selected must be comparable to certified toxin standards that are commercially available from the National Research Council Halifax, Canada. The final step in this analysis is to calculate the net toxicity (expressed in g STXeq) from the molar specific potencies (MU mol^{-1}) of individual PSP toxins (Oshima 1995).

A post-column derivatization method for PST analysis was developed in the late 1970s and involved using a silica-based stationary phase (Buckley et al. 1978). Post-column derivatization allows continuous analysis (via post-column oxidation) of toxins and subsequent detection by fluorescence. The toxins are separated on an ion exchange column (Oshima et al. 1989; Oshima 1995). This method is valued for producing results that correlate well to those produced by MBA. A post-column method for analysis of PSTs was standardized by a biotoxin working group of the European Committee for Standardization (CEN) and was approved as the European Norm EN 14194 (CEN 2002). However, a pre-column derivatization approach (employing a single reverse phase column) was developed, in which peroxide or periodate was used as the oxidation agent (Lawrence et al. 1991). For instrumental analyses, the pre-column approach is much simpler than the post-column approach, and this pre-column method was approved by the Association of Official Analytical Chemists (AOAC) in June 2005. This so-called pre-column derivatization liquid chromatography has high accuracy and precision, and produces data that are significantly correlated to MBA data.

Generally, pre-column derivatization is more sensitive for detecting lower toxin levels than the post-column derivatization method. However, the pre-column method provides a total toxin value, because the following toxins co-elute: GTX2 and 3, GTX1 and 4 and dcGTX2 and dcGTX3. Therefore, the post-column oxidation method is more accurate for individual PSP toxin analysis (Rodríguez et al. 2010). Nevertheless, HPLC methods of PSP toxin analysis are slow, costly and technically demanding. For these reasons, HPLC methods are usually carried out only in centralized laboratories.

5.2.2 Liquid Chromatography-Mass Spectrometry (LC-MS)

Liquid chromatography-mass spectrometry (LC-MS) is a technique that allows identification of unknown compounds, elucidation of structural properties of molecules and quantification of known compounds. LC-MS has been advocated by Quilliam (1998) as a universal analytical method for all marine toxins (Humpage et al. 2010). Its main advantages are high specificity, high sensitivity, rapid analysis, automation, elimination of false positives, and limited sample preparation, compared to other methods, which require complex derivatization and cleanup protocols. However, the main drawbacks of this technique are initial high capital equipment cost, the requirement for having highly skilled technicians, and the need for calibration standards (which are not available for all biotoxins).

5.3 Immunosorbant Assays

Immunoassays are biochemical assays that utilize antibodies for analysis. Antibodies obtained from a conjugate with STX shows the highest sensitivity for STX, dcSTX and GTX2, 3, whereas, antibodies obtained from a conjugate with neoSTX has high sensitivity for neoSTX and GTX1, 4 (Usleber et al. 2001). A general quantitative immunoassay specific to all STX group toxins is not available (by using only one type of antibody). Therefore, immunoassays were developed and are used for pre-screening analyses, which reduces the requirement for mouse bioassays. Fortunately, immunoassay provides qualitative (presence/absence) screening results for saxitoxin in less than 20 min (Inami et al. 2004).

A range of immunoassays have been developed with poly- and mono-clonal antibodies to analyze for saxitoxin and several of its derivatives. Commercially available kits include the Ridascreen fast saxitoxin test (R-BioPharm), the Abraxis ELISA for PSP (Abraxis) and the Maxsignal saxitoxin ELISA (Bio Scientific). The Ridascreen assay, with monoclonal antibodies, has a much better (0.02 µg L^{-1}) detection limit than does the polyclonal antibody-based (Abraxis and MaxSignal) assays, the limit of detection for which is 1.2 µg L^{-1} (Usleber et al. 2001). The immunoassay allows rapid screening for detecting positive samples and is very helpful to regulatory agencies, shellfish processing plants and the aquaculture industry. The positive samples that are detected by using this assay can be subjected to further quantitative analysis by using more definitive methods. One disadvantage of these immunoassays is the differential affinity of the antibody mixture for individual PSP toxins.

6   Biotransformation of PSTs

Most PSTs in marine bivalves are biotransformed in the digestive gland, indicating the presence therein of toxin transforming enzymes or bacteria (Fast et al. 2006; Lu and Hwang 2002). The concentration of toxins in bivalves is generally similar to the concentration in the dinoflagellates on which they feed. Bivalves, however, usually contain higher amounts of the carbamate toxin (GTXs) form than do the dinoflagellates, and lower levels of the N-sulfocarbamoyl (C) toxins (Kwong et al. 2006; Choi et al. 2003). This difference appears to result from the biotransformation of these PSTs by the bivalves from the less stable β-epimer (C2, GTX3, GTX4) form to the more stable α-epimer (C1, GTX1, GTX2) form, until these moieties reach an α:β ratio of 1:3 (Bricelj and Shumway 1998; Oshima 1995). The transformation of the C2 toxin to GTXs and dcGTX increases the net toxicity of the shellfish to consumers by four to tenfold. This transformation process is accelerated at higher temperature and pH (Oshima 1995). The biotransformation of toxins possibly involves both enzymatic and non-enzymatic conversion reactions. The probable metabolic mechanisms involved include desulfation, oxidation, reduction and epimerization. Such transformation mechanisms are known to alter the overall toxicity of the PSTs.

Kotaki et al. (1985) were the first to suggest that the PSTs are biotransformed by the bacteria, *Vibrio* and *Pseudomonas* spp. Most of the early studies demonstrated that bacteria transform less toxic PSTs to more toxic PSTs. Conversion of GTX5 to STX was documented to occur in Blue mussels (Sullivan 1982). Moreover, bacterial isolates from the viscera of marine crabs, snails and red algae have also been demonstrated to transform GTX 1-4 to STX and neoSTX (Sugawara et al. 1997; Kotaki 1989; Kotaki et al. 1985). Transformation of these less toxic PSTs (e.g., GTX 1, 2 and 3) to more toxic ones (e.g., STX and neoSTX) is associated with the reductive elimination of the O_{22}-sulfate and N1-hydroxyl groups (Kotaki et al. 1985; Shimizu and Yoshioka 1981). Nevertheless, the reaction only occurs after other carbon sources are exhausted (Jones et al. 1994). The reaction was faster under anaerobic than under aerobic conditions (Kotaki et al. 1985).

Enzymatic transformation of more toxic PSTs (carbamoyl and carbamoyl-N sulfated toxins) to less PSTs (decarbamoyl compounds) has been documented to occur in green mussels (Choi et al. 2003), and in littleneck clams and scallops (Noguchi et al. 1989). Decarbamoylase clearage of the carbamoyl group in carbamoyl and carbamoyl-N sulphated toxins can result in the formation of significantly less toxic decarbamoyl toxins (Smith et al. 2001). Bacterial sulfotansferases also transformed the more toxic PST, GTX 2/3 and GTX 1/4, into less toxic PSTs, C1/C2 and C3/C4, respectively, via enzymatic oxidation (Lee et al. 1995). Biotransformation of STX, GTX 2 and GTX 3 into GTX 5, C1 and C2, respectively, by sulfotransferase activity has been described for the dinoflagellates *G. catenatum* and *Alexandrium catenella* (Yoshida et al. 1998). Moreover, bacterial isolates from bivalve gut have been shown to degrade GTX1 and GTX4, without the appearance of GTX2 and GXT3 (Smith et al. 2001). *Pseudoalteromonas haloplanktis*, isolated from the digestive tracts of blue mussels, has been shown to play an important role in reducing the toxicity of PST by 90% within 3 days (Donovan et al. 2009).

Non-enzymatic transformation (detoxification by chemical agents) by ozonation is commonly used to inactivate toxins from extracts of dinoflagellates and shellfish tissues. Ozone treatment is more effective than chlorine treatment for depuration of PSP in green mussels and clams. Ozone treatment completely detoxified the toxins in bivalves within 10 days, while chlorine treatment required 15 days to achieve similar results. Ozone treatment is preferred, because it inactivates bacteria or virus, oxidizes toxins, and does not change the taste and appearance of the shellfish (Sharma et al. 2011). Nevertheless, ozone treatment does not completely remove toxins that are incorporated into tissues. Detoxification is influenced by the diet of bivalves during the depuration period. For example, oysters that feed on a mixture of *Chlorella* and chitosan demonstrated higher depuration efficiency (decrease from 12 MU g^{-1} into 0.5 MU g^{-1} in 7 days) than those feeding only on *Chlorella* or chitosan (decreased from 12 MU g^{-1} into 3.47 MU g^{-1} and 1.4 MU g^{-1}, respectively, in 7 days) (Xie et al. 2013). This is because active feeding is likely to accelerate gut evacuation rates and overall metabolism (degradation, excretion) of the toxins. In addition, shellfish have the natural ability to chemically transform GTX1/4 to GTX2/3 by reduction of the N1-OH group in GTX1 and GTX4 to an N1-H group from conjugation with glutathione and cysteine (Oshima 1995).

7 Conclusions

Paralytic shellfish poisoning is a global problem that affects coastal communities throughout the world. The causative toxins, saxitoxins and saxitoxin-related compounds are potent neurotoxins that negatively affect the fisheries industry and human health. In general, nutrient concentration, the nutrient ratio and environmental parameters are known to influence harmful algal blooms (HABs) and the cellular toxicity they cause. However, the linkage among nutrients (organic and inorganic), environmental parameters, PST toxicity and PST composition are still poorly understood. Further investigations are needed to elucidate how certain environmental parameters, certain nutrients and their proportions act to regulate the production of PSTs in HAB species.

Filter feeding bivalves are the primary vector of paralytic shellfish poisoning. The bioaccumulation of PSTs in bivalves is associated with shellfish sensitivity to the toxins. Sensitive shellfish usually accumulate lower toxin levels, while insensitive species accumulate toxins to an extremely dangerous level. To ensure the safety of seafood products, shellfish fisheries are closed by regulators if any edible portion of shellfish is found to have levels of PSTs equal to or exceeding 0.8 μg STXeq per 100g tissue shellfish flesh. The closures of such facilities are often long and sometimes indefinite. Therefore, for the benefit of processors, it is important that scientists understand the mechanism and rate of uptake of PSTs, as well as their elimination rate from different bivalve species. Understanding the kinetics of uptake and elimination will enhance our ability to balance protecting public health, while preserving the economic viability of the seafood industry.

There is clear evidence that PSTs are biotransformed by bacterial isolates from bivalves. However, a vast knowledge gap exists about the enzymatic pathways of how the PSTs are biotransformed to less toxic versions of themselves by the bivalves or by the bacteria that infest the bivalves. In addition, little is known of the factors, either environmental or otherwise that affect PST biotransformation. In particular, research is needed to elucidate the factors that influence both non-enzymatic and enzymatic transformation of PST by bacteria. Specific bacterial strains should be examined that can be utilized in the bivalve culture industry or elsewhere to degrade or detoxify PSTs. Development of novel methods of detoxification is essential from both the consumer human health and industry economic viability perspectives.

8 Summary

Saxitoxin and related compounds are potent marine neurotoxins, and are often associated with paralytic shellfish poisoning (PSP) incidents. In marine waters, dinoflagellates belonging to the genera *Alexandrium*, *Gymnodinium* and *Pyrodinium* are the agents that cause PSP. The growth rate, PST (paralytic shellfish toxin) concentration and composition produced by these PST-generating dinoflagellates are highly influenced by dietary nutrient concentration, nutrient composition and other environmental factors. With the exception of filter-feeding bivalves, all marine organisms are vulnerable to the effects of the PSTs. The suspension-feeding bivalves are the primary vectors for transferring the PSTs that pose a health hazard to humans and other food web consumers. Different bivalve species produce different levels of PSTs. The toxin levels in these bivalves species depends mainly on how much of the toxins they accumulate (which corresponds to their toxin sensitivity), and on the rate at which they depurate the accumulated toxin. Because of the toxic effects they produce in consumers, the presence of PSTs is monitored in shellfish and sometimes in environmental samples. In general, the analysis of PSTs is primarily achieved by utilizing bioassays (i.e., *in vivo, in vitro* cell assay), *in vitro* functional assays, immunosorbant assays or/and by chemical analysis.

The results of many studies have provided clear evidence that both bivalves and bacteria have the ability to biotransform and detoxify PSTs. A major future research goal should be to improve the understanding of how PSTs are degraded and biotransformed into less toxic analogs. Gaining such information will help to develop new approaches for detoxifying PST contaminated shellfish that are destined for human consumption.

Acknowledgements This work was financially supported by the Niche Research Grant Scheme (NRGS0003) from the Ministry of Education Malaysia.

References

Acres J, Gray J (1978) Paralytic shellfish poisoning. Can Med Assoc J 119:1195–1197
Al-Azad S, Tan KS, Ransangan J (2013) Effects of light intensities and photoperiods on growth and proteolytic activity in purple non-sulfur marine bacterium, *Afifella marina* strain ME (KC205142). Adv Biosci Biotechnol 4:919–924
Anderson DM (2009) Approaches to monitoring, control and management of harmful algal blooms (HABs). Ocean Coast Manag 52:342
Anderson DM (1989) Toxic algal blooms and red tides: a global perspective. In: Okaido AT, Anderson DM, Nemoto T (eds) Red tides: biology environmental science and toxicology. Elsevier, New York, NY, pp 11–16
Anderson DM, Kulis DM, Sullivan JJ, Hall S (1990) Toxin composition variations in one isolate of the dinoflagellate Alexandrium fundyense. Toxicon 28:885–893
Anderson DM, Glibert PM, Burkholder JM (2002) Harmful algal blooms and eutrophication: nutrient sources, composition, and consequences. Estuaries 25:704–726
Anton A, Alexander J, Estim A (2000) Harmful algal blooms in Malaysia: revisiting Kimanis Bay. In: 9th International Conference on Toxic Phytoplankton, Tasmania (Abstract)
AOAC International (1995) Paralytic shellfish poison. Biological method. Chapter 35. In: Williams S (ed) Official methods of analysis, 14th edn. Association of Official Analytical Chemists International, Arlington, VA, pp 21–22
Aune T, Ramstad H, Heidenreich B, Landsverk T, Waaler T, Egaas E, Julshamn K (1998) Zinc accumulation in oysters giving mouse deaths in paralytic shellfish poisoning bioassay. J Shellfish Res 17:1243–1246
Aune T, Aasen JAB, Miles CO, Larsen S (2008) Effect of mouse strain and gender on LD50 of yessotoxin. Toxicon 52:535–540
Baden DG, Trainer VL (1993) Mode of action of saxitoxins of seafood poisoning. In: Falconer I (ed) Algal toxin in seafood and drinking water. Academic, London, pp 49–74
Beitler MK, Liston J (1990) Uptake and tissue distribution od PSP toxin in butter clams. In: Graneli E, Sindstrom B, Edler L, Anderson DM (eds) Toxic marine phytoplankton. Elsevier Science Publications, Amsterdam, pp 257–263

Benavides H, Prado L, Díaz S, Carreto JI (1995) An exceptional bloom of *Alexandrium catenella* in the Beagle Channel, Argentina. In: Lassus P, Arzul G, Erard-Le Denn E, Gentien P, Marcaillou-Le Baut C (eds) Harmful marine algal blooms. Lavoisier Publishers, Paris, pp 113–119

Bricelj VM, Lee JH, Cembella AD, Anderson DM (1990) Uptake kinetics of paralytic shellfish toxins from the dinoflagellate *Alexandrium fundyense* in the mussel *Mytilus edulis*. Mar Ecol Prog Ser 63:177–188

Bricelj VM, Lee JH, Cembella AD (1991) Influence of dinoflagellate cell toxicity on uptake and loss of paralytic shellfish toxins in the northern quahog, *Mercenaria mercenaria*. Mar Ecol Prog Ser 74:33–46

Bricelj VM, Cembella AD (1995) Fate of gonyautoxins accumulated in surfclams, Spisula solidissima, grazing upon PSP toxin-producing Alexandrium. In: Lassus P, Arzul G, Erard E, Gentien P, Marcaillou C (eds) Harmful marine algal blooms. Lavoisier Science Publishers, Paris

Bricelj VM, Cembella AD, Laby D, Shumway SE, Cucci TL (1996) Comparative physiological and behavioral responses to PSP toxins in two bivalve molluscs, the softshell clam, Mya arenaria, and surfclam, Spisula solidissima. In: Yasumoto T, Oshima Y, Fukuyo Y (eds) Harmful and toxic algal blooms. Intergovernmental Oceanographic Commission of UNESCO, Paris, pp 405–408

Bricelj VM, Shumway SE (1998) Paralytic shellfish toxins in bivalve molluscs: occurrence, transfer kinetics and biotransformation. Rev Fisher Sci 6(4):315–383

Brivelj VM, Connell L, Konoki K, Macquarrie SP, Scheuer T, Catteral WA, Trainer VL (2005) Sodium channel mutation leading to saxitoxin resistance in clams increases risk of PSP. Nature 434:764–767

Buckley LJ, Oshima T, Shimizu Y (1978) Construction of a paralytic shellfish toxin analyzer and its application. Anal Biochem 85:157–164

Carreto JI, Benavides HR, Negri RM, Glorioso PD (1986) Toxic red-tide in the Argentine Sea. Phytoplankton distribution and survival of the toxic dinoflagellate Gonyaulax excavata in a frontal area. J Plankton Res 8:15–28

CEN (2002) EN 14194. Foodstuffs - determination of saxitoxin and desaxitoxin in mussels - HPLC method using post-column derivatization with peroxide or periodate oxidation. European Committee for Standardization (CEN).

Cestele S, Catterall WA (2000) Molecular mechanisms of neurotoxin action on voltage-gated sodium channels. Biochemistry 82:883–892

Chen CY, Chou HN (2001) Accumulation and depuration of paralytic shellfish toxins by purple clam Hiatula rostrata. Light Tools Toxicon 39:1029–1034

Choi MC, Hsieh DPH, Lam PKS, Wang WX (2003) Field depuration and biotransformation of paralytic shellfish toxins in scallop Chlamys nobilis and green- lipped mussel Perna viridis. Mar Biol 143:927–934

Chou HN, Huang CP, Chen CY (2005) Accumulation and depuration of paralytic shellfish poisoning toxins by laboratory cultured purple clam Hiatula diphos Linnaeus. Toxicon 46:587–590

Ciminiello P, Fattorusso E, Forino M, Montresor M (2000) Saxitoxin and neosaxitoxin as toxic principles of Alexandrium Anderson (Dinophyceae) from the Gulf of Naples, Italy. Toxicon 38:1871–1877

Cordier S, Monfort C, Miossec L, Richardson S, Belin C (2000) Ecological analysis of digestive cancer mortality related to contamination by diarrhetic shellfish poisoning toxins along the coast of France. Environ Res 84:145–150

Dahl E, Tangen K (1993) 25 years experience with *Gyrodinium aureolum* in Norwegian waters. In: Smayda TJ, Shimizu Y (eds) Toxic phytoplankton blooms in the sea. Elsevier, New York, NY, pp 15–21

Daigo K, Noguchi P, Miwa A, Kawai N, Hasimoto K (1988) Resistance of nerves from certain toxic crabs to paralytic shellfish poison and tetrodotoxin. Toxicon 26:485–490

Davio SR, Fontelo PAA (1984) Competitive displacement assay to detect saxitoxin and tetrodotoxin. Anal Biochem 141:199–204

Deeds J, Landsberg J, Etheridge S, Pitcher G, Longan S (2008) Non-traditional vectors for paralytic shellfish poisoning. Mar Drugs 6:308–348

Dell'Aversano C, Walter JA, Burton IW, Stirling DJ, Fattorusso E, Quilliam MA (2008) Isolation and structure elucidation of new and unusual saxitoxin analogues from mussels. J Nat Prod 71:1518–1523

Deyoe HR, Sutrle CA (1994) The inability of the Texas "brown tide" alga to use nitrate and the role of nitrogen in the initiation of a persistent Moom of tiffs organism. J Physiol 7(30):800–806

Donovan CJ, Ku JC, Quiliam MA, Gill TA (2008) Bacterial degradation of paralytic shellfish toxins. Toxicon 52:91–100

Donovan CJ, Garduno RA, Kalmokoff M, Ku JC, Quilliam MA, Gill TA (2009) Pseudoalteromonas bacteria are capable of degrading paralytic shellfish toxins. Appl Environ Microbiol 75:6919–6923

Doucette GJ, Logan MM, Ramsdell JS, van Dolah FM (1997) Development and preliminary validation of a microtiter plate-based receptor binding assay for paralytic shellfish poisoning toxins. Toxicon 35:625–636

Fast MD, Cembella AD, Ross NW (2006) In vitro transformation of paralytic shellfish toxins in the clams Mya arenaria and Protothaca staminea. Harmful Algae 5:79–90

Franco J, Fernandez P, Reguera B (1994) Toxin profiles of natural populations and cultures of Alexandrium minitum Halim from Galician (Spain) coastal waters. J Appl Phycol 6:275–279

Gacutan RQ, Tabbu MY, de Castro T, Gallego AB, Arafiles MB, Icatlo F (1989) Detoxification of Pyrodinium generated paralytic shellfish poisoning toxin in Perna viridis from western Samar, Philippines. In: Hallegraeff GM, Maclean JL (eds), Biology, epidemiology and management of Pyrodinium red tides. JCLARM conference proceeding 21. Manila, Philippines, pp 80–85.

Gainey LF Jr, Shumway SE (1988) A compendium of the responses of bivalve molluscs to toxic dinoflagellates. J Shellfish Res 7(4):623–628

Garate-Lizarraga I, Bustillos-Guzman JJ, Morquecho L, Band-Schmidt CJ, Alonso-Rodriguez R, Erler K, Luckas B, Reyes-Salinas A, Gongora-Gonzalez DT (2005) Comparative paralytic shellfish poisoning profiles in the strains of Gymnodinium catenatum Graham from the Gulf of California, Mexico. Mar Pollut Bull 50:211–217

Glasgow HB, Burkholder JM, Mallin MA, Dreamer-Melia NJ, Reed RE (2001) Field ecology of toxic *Pfiesteria* complex species, and a conservative analysis of their role in estuarine fish kills. Environ Health Perspect 109:715–730

Glibert PM, Magnien R, Lomas MW, Alexander J, Fan C, Haramoro E, Trice M, Kana TM (2001) Harmful algal blooms in the Chesapeake and coastal bays of Maryland, USA: comparison of 1997, 1998, and 1999 events. Estuaries 24:875–883

Graneli E, Carlsson P (1998) The ecological significance of phagoix-ophy in photosynthetic flagellates. In: Anderson DM, Cerebella AD, HallegTaeff GM (eds) Physiological ecology of harmful algal blooms. Springer, Berlin, Germany, pp 540–557

Grzebyk D, Bechemin C, Ward CJ, Verite C, Codd GA, Maestrini SY (2003) Effects of salinity and two coastal waters on the growth and toxin content of the dinoflagellate *Alexandrium minitum*. J Plankton Res 25:1185–1199

Hamasaki K, Horie M, Tokimitsu S, Toda T, Taguchi S (2001) Variability in toxicity of the dinoflagellate *Alexandrium tamarense* isolated from Hiroshima Bay, western Japan, as a reflection of changing environmental conditions. J Plankton Res 23:271–278

Hallegraeff GM (1993) A review of harmful algal blooms in the Australian region. Mar Pollut Bull 25:186–190

Hansen PJ, Cembella AD, Moestrup O (1992) The marine dinoflagellate Alexandrium ostenfeldii: paralytic shellfish toxin concentration, composition and toxicity to a tintinid ciliate. J Phycol 28:597–603

Hodgkiss IJ, Ho KC (1997) Are changes in N:P ratios in coastal waters the key to increased red tide blooms. Hydrobiologia 352:141–147

Holmes MJ, Bolch CJS, Green DH, Cembella AD, Teo SLM (2002) Singapore isolates of dinoflagellate Gymnodinium catenatum (Dinophyceae) produce a unique profile of paralytic shellfish poisoning toxins. J Phycol 38:96–106

Humpage AR, Magalhaes VF, Froscio SM (2010) Comparison of analytical tools and biological assays for detection of paralytic shellfish poisoning toxins. Anal Bioanal Chem 397:1655–1671

Hurst JW, Gilfillan ES (1997) Paralytic shellfish poisoning in Maine. In: Wilt ES (ed) Tenth Natl shellfish sanitation workshop. U.S. Dept. Health, Education and Welfare, Food and Drug Administration, Washington, DC, pp 152–161

Hwang DF, Lu YH, Noguchi T (2003) Effects of exogenous polyamines on growth, toxicity, and toxin profile of dinoflagellate Alexandrium minutum. J Food Hygien Soc Jpn 44:49–53

Hwang DR, Lu YH (2000) Influence of environmental and nutritional factors on growth, toxicity, and toxin profile of dinoflagellate Alexandrium minitum. Toxicon 38:1491–1503

Ichimi K, Suzuki T, Ito A (2002) Variety of PSP toxin profile in various culture strains of Alexandrium tamarense and change of toxin profile in natural A. tamarense population. J Exp Mar Biol Ecol 273:51–60

Inami GB, Crandall C, Csuti D, Oshiro M, Brenden RA (2004) Feasibility of reduction in use of mouse bioassay: presence/absence screening for saxitoxin in frozen acidified mussel and oyster extracts from the coast of California with in vitro methods. J AOAC Int 87(5):1133–1142

Jacobson DM, Anderson DM (1996) Widespread phagocytosis of dilates and other protists by marine mixotrophic and heterotrophic thecate dinoflagellates. J Phycol 82:279–285

Jaime E, Gerdts G, Luckas B (2007) In vitro transformation of PSP toxins by different shellfish tissues. Harmful Algae 6:308–316

Jellett JF, Marks LJ, Stewart JE, Dorey MI, Watson-Wright W, Lawrence JF (1992) Paralytic shellfish poisoning (saxitoxin family) bioassays: automated endpoint determination and standardization of the in vitro tissue culture bioassay, and comparison with the standard mouse bioassay. Toxicon 30:1143–1156

Jellett JF, Doucette LI, Belland ER (1998) The MIST ™ shippable cell bioassay kids for PSP: an alternative to the mouse bioassay. J Shellfish Res 17:1653–1655

Johansson N, Graneli E (1999) Cell density, chemical composition and toxicity of *Chrysochromulina polylepis* (Haptophyta) in relation to different N:P supply ratios. Mar Biol 135:209–217

Johansson N, Graneli E, Yasumoto T, Carlsson P, Legrand C (1996) Toxin production by *Dinophysis acuminata* and *D. acuta* cells grown under nutrient sufficient and deficient conditions. In: Yasumoto T, Oshima Y, Fukuyo Y (eds) Harmful and toxic algal blooms. Scientific and Cultural Organization, Intergovernmental Oceanographic Commission of United Nations Educational, Paris, France, pp 227–280

Jones GJ, Bourne DG, Blakeley RL, Doelle H (1994) Degradation of the cyanobacterial hepatotoxin microcystin by aquatic bacteria. Nat Toxins 2:228–235

Kao CY (1993) Paralytic shellfish poisoning. Algal toxins in seafood and drinking water. Academic, London, pp 75–86

Kodama M (1990) Possible link between bacteria and toxin production in algal blooms. In: Granelii EP, Sundstrom B, Edler L, Anderson DN (eds) Toxic marine phytoplankton. Elsevier, New York, NY, pp 52–61

Kodama M (2000) Ecobiology, classification, and origin. In: Botana LM (ed) Seafood and freshwater toxins - pharmacology, physiology, and detection. Marcel Dekker, New York, NY, pp 125–149

Kotaki Y, Oshima Y, Yasumoto T (1985) Bacterial transformation of paralytic shellfish toxins in coral reef crabs and marine snail. Nippon Suisan Gakkashi 51:1009–1013

Kotaki Y (1989) Screening of bacteria which convert gonyautoxin 2, 3 to saxitoxin. Nippon Suisan Gakkashi 55:1293

Krock B, Seguel CG, Gembella AD (2007) Toxin profile of Alexandrium catenella from the Chilean coast as determined by liquid chromatography with fluorescence detection and liquid chromatography coupled with tandem mass spectrometry. Harmful Algal 6:734–744

Kwong RW, Wang WX, Lam PK, Yu PK (2006) The uptake, distribution and elimination of paralytic shellfish toxins in mussels and fish exposed to toxic dinoflagellates. Aquat Toxicol 80:82–91

Larocque R, Cembella AD (1991) Résultats du premier programme de suivi des populations de phytoplancton toxique dans l'estuaire et le Golfe du Saint-Laurent (Région du Québec). Rapp Tech Can Sci Hal Aquat 1796: 42 p

Lassus P, Ledoax M, Bardouill M, Bohee M (2000) Comparative efficiencies of different non-toxic microalgal diet in detoxification of PSP-contaminated oyster. J Nat Toxin 9:1–12

Lawrence JF, Menard C, Charbonneau CF, Hall S (1991) A study of ten toxins associated with paralytic shellfish poison using prechromatographic oxidation and liquid chromatography with fluorescence detection. J AOAC 74:404–409

Lee NS, Kim BT, Kim DH, Kobashi K (1995) Purification and reaction mechanism of arylsulfate sulfotransferase from *Haemophilus* K-12, a mouse intestinal bacterium. J Biochem 118:796–801

Lefebvre KA, Bill BD, Erickson A, Baugh KA, O'Rourke L, Costa PR, Nance S, Trainer VL (2008) Characterization of intracellular and extracellular saxitoxin levels in both field and cultured Alexandrium spp. Samples from Sequim Bay, Washington. Mar Drugs 6:103–116

Lewitus AJ, Haves KC, Gransden SG, Glascow HB, Burkholder JM Jr, Glibert PM, Morton SL (2001) Ecological characterization of a widespread *Scrippsiella* red tide in South Carolina estuaries: a newly observed phenomenon. In: Hallegraeff GM, Blackbuna S, Bolch C, Lewis R (eds) Proceedings of the ninth international conference on harmful algal blooms. Intergovernmental Oceanographic Commission, United Nations Educational Scientific, and Oalmral Organization, Paris, France

Li A, Stoeker DK, Coats DW (2000) Spatial and temporal aspects of *Gyrodinium galatheanum* in Chesapeake Bay: distribution and mixotrophy. J Plankton Res 22:2105–2124

Li A, Stoeker DK, Coats DW (2001) Mixotrophy in *Gyrodinium galatheanum* (Dinophyceae): grazing responses to light intensity and inorganic nutrients. J Phycol 36:33–45

Lim PT, Leaw CP, Usup G (2001) First incidence of paralytic shellfish poisoning on the east coast of Peninsular Malaysia. In: Sasekumar A, Usup G, Noraieni M, Ung EH, Lee SC (eds), Book of abstracts Asia-Pacific Conference on marine science & technology. Marine science into the new millennium: new perspectives & challenges, 12–16 May 2001, Kuala Lumpur, Malaysia

Lim PT, Usup G, Leaw CP, Ogata T (2005) First report of Alexandrium taylori and Alexandrium peruvianum (Dinophyceae) in Malaysia waters. Harmful Algae 4:391–400

Lim PT, Ogata T (2005) Salinity effect on growth and toxin production of four tropical *Alexandrium* species (Dinophyceae). Toxicon 45:699–710

Lim PT, Leaw CP, Usup G, Kobiyama A, Koike K, Ogata T (2006) Effects of light and temperature on growth, nitrate uptake, and toxin production of two tropical dinoflagellates: *Alexandrium tamiyavanichi* and *Alexandrium minutum* (Dinophyceae). J Phycol 42:786–799

Lim PT, Leaw CP, Kobiyama A, Ogata T (2010) Growth and toxin production of tropical *Alexandrium minutum* Halim (Dinophyceae) under various nitrogen to phosphorus ratios. J Appl Phycol 22:203–210

Lippemeier S, Frampton DMF, Blackburn SI, Geier SC, Negri AP (2003) Influence of phosphorus limitation on toxicity and photosynthesis of Alexandrium minutum (Dinophyceae) monitored by in-line detection of variable chlorophyll fluorescence. J Phycol 39(2):320–331

Llewellyn LE (2006) Saxitoxin, a toxic marine natural product that targets a multitude of receptors. Nat Prod Rep 23:200–222

Lomas MW, Glibert EM (2000) Comparisons of nitrate uptake, storage, and reduction in *marine* diatoms and flagellates. J Phycol 36:903–913

Lu YH, Hwang DF (2002) Effects of toxic dinoflagellates and toxin biotransformation in bivalves. J Nat Toxins 11:315–322

Mahar J, Lukàcs GL, Li Y, Hall S, Moczydlowski E (1991) Pharmacological and biochemical properties of saxiphilin, a soluble saxitoxin-binding protein from the bullfrog (*Rana catesbiana*). Toxicon 29:53–71

Martin JL, White AW, Sullivan JJ (1990) Anatomical distribution of paralytic shellfish toxins in softshell clams. In: Granéli E, Sundström B, Edler L, Anderson DM (eds) Toxic marine phytoplankton. Elsevier, New York, NY, pp 379–384

Negri AP, Bolch CJS, Geier S, Green DH, Park TG, Blackburn SI (2007) Widespread present of hydrophobic paralytic shellfish toxins in Gymnodinium catenatum. Harmful Algae 6:774–780

Negri AP, Jones GJ (1995) Bioaccumulation of paralytic shellfish poisoning (PSP) toxins from the cyanobacterium Anabaena circinalis by the freshwater mussel Alathyria condola. Toxicon 33:667–678

Noguchi T, Chen S, Arakawa O, Hashimoto K (1989) A unique composition of PSP in "hoigi" scallop Chlamys nobilis. In: Natori S, Hashimoto K, Ueno Y (eds) Mycotoxins and phycotoxins'88. Elsevier, Amsterdam, pp 351–358

Nygaard K, Tobiesen A (1998) Bacterivory in algae: a survival strategy during nutrient limitation. Limnol Oceanogr 38:273–279

Ogata T, Kodama M, Ishimaru T (1989) Effect of water temperature and light intensity on growth rate and toxin production in toxic dinoflagellates. In: Okaichi T, Anderson DM, Nemoto T (eds) Red tides: biology, Environmental science and toxicology. Elsevier, New York, NY, pp 423–426

Ogata T, Pholpunthin P, Fukuyo Y, Kodama M (1990) Occurrence of *Alexandrium cohorticula* in Japanese coastal water. J Appl Phycol 2:351–356

Oshima Y, Sugino K, Yasumoto T (1989) Latest advance in HPLC analysis of paralytic shellfish toxins. In: Natori S, Hashimoto K, Ueno Y (eds) Mycotoxins and phycotoxins'88. Elsevier Science Publishers, Amsterdam, pp 319–326

Oshima Y, Blackburn SI, Hallegraeff GM (1993) Comparative study on paralytic shellfish toxin profiles of the dinoflagellates Gymnodinium catenatum from three different countries. Mar Biol 116:471–476

Oshima Y (1995) Post-column derivatization HPLC method for the analysis of PSP. J AOAC Int 78:795–799

Parkhill J, Cembella A (1999) Effects of salinity, light and organic nitrogen on growth and toxigenicity of marine dinoflagellate Alexandrium tamarense from northeastern Canada. J Plankton Res 21:939–955

Pitcher GC, Cembella AD, Joyce LB, Larsen J, Probyn TA, Ruiz Sebastian C (2007) The dinoflagellate Alexandrium minutum in Cape Town harbour (South Africa): bloom characteristics, phylogenetic analysis and toxin composition. Harmful Algae 6:823–836

Poulton NJ, Keafer BA, Anderson DM (2005) Toxin variability in natural population of Alexandrium fundyense in Casco Bay, Maine-evidence of nitrogen limitation. Deep-Sea Res 52(PT2):2501–2521

Proctor NH, Chan SL, Truvor AJ (1975) Production of saxitoxin by culture of Gonyaulax catenella. Toxicon 13:1–19

Quilliam MA, Janecek M (1993) Characterization of oxidation products of paralytic shellfish poisoning toxins by liquid chromatography/mass spectrometry. Rapid Comm Mass Spectrom 7:482–487

Quilliam MA (1998) Phycotoxins. J AOAC Int 81:142–151

Rodrigue DC, Etzel RA, Hall S, de Porras E, Velasquez OH, Tauxe RV, Kilbourne EM, Blake PA (1990) Lethal paralytic shellfish poisoning in Guatemala. Am J Trop Med Hyg 42:267–271

Rodríguez P, Alfonso A, Botana AM, Vieytes MR, Botana LM (2010) Comparative analysis of pre- and post-column oxidation methods for detection of paralytic shellfish toxins. Toxicon 56:448–457

Romdhane MS, Eilertsen HC, Yahia OKD, Yahia MND (1998) Toxic dinoflagellate blooms in Tunisian lagoons: causes and consequences for aquaculture. In: Reguera B, Blance J, Fernandez ML, Wyatt T (eds) Harmful algae. Xunta de Galicia and Intergovernmental Oceanographic Commission of United Nations Educational, Scientific and Cultural Organization, Paris, France, pp 80–83

Schantz EJ, McFarren EF, Schafer ML, Lewis KH (1958) Purified shellfish poison for bioassay standardization. J Assoc Off Anal Chem 41:160–168

Schantz EJ, Ghazarossian VE, Schnoes HK, Strong FM, Springer JP, Pezzanite JO, Clardy J (1975) Structure of saxitoxin. J Am Chem Soc 97:1238–1239

Sebastian CR, Etheridge SM, Cook PA, O'Ryan C, Pitcher GC (2005) Phylogenetic analysis of toxic Alexandrium (Dinophyceae) isolates from South Africa: implications for the global phylogeography of Alexandrium tamarense species complex. Phycologia 44:49–60

Sharma R, Venkateshvaran K, Purushothaman CS (2011) Bioaccumulation and depuration of paralytic shellfish toxin in *Perna viridis* and *Meretrix meretrix* from Mumbai, India. Ind J Marine Sci 40:542–549

Shimizu Y, Yoshioka M (1981) Transformation of paralytic shellfish toxins as demonstrated in scallop homogenates. Science 212:547–549

Shimizu Y, Watanabiz N, Wrensfori G (1993) Biosynthesis of brevetoxins and heterotrophic metabolism in *Gymnodinium breve*. In: Lassus R, Arzul C, Erard-Le-Denn E, Gentian R, Mm-caillou C (eds) Harmful marine algal blooms. Lavoisier Publishing, Paris, France, pp 351–357

Silvert W, Bricelj M, Cembella A (1998) Dynamic modelling of PSP toxicity in the surfcalm (Spisula solidissima): multicompartmental kinetics and biotransformation. In: Rguera B, Blanco J, Fernandez ML, Wyatt T (eds) Harmful algae. VIII international Conference. Intergovernmental Oceanographic Commission of UNESCO, Paris, pp 437–440

Siu G, Young M, Chan D (1997) Environmental and nutritional factors which regulate population dynamics and toxin production in the dinoflagellate Alexandrium catenella. Hydrobiology 352:117–140

Smil V (2001) Enriching the earth: Fritz Haber, Carl Bosch, and the transformation of world food. Tile MIT Press, Cambridge, UK

Smith EA, Grant F, Ferguson CMJ, Gallacher S (2001) Biotransformations of paralytic shellfish toxins by bacteria isolated from bivalve molluscs. Appl Environ Microbiol 67(5):2345–2353

Sommer H, Meyer KF (1937) Paralytic shellfish poisoning. Achiev Pathol Labor Med 24: 560–598

Soon TK, Ransangan J (2014) A review of feeding behavior, growth, reproduction and aquaculture site selection for green-lipped mussel, Perna viridis. Adv Biosci Biotechnol 5:462–469

Su Z, Sheets M, Ishida H, Li FH, Barry WH (2004) Saxitoxin blocks L-type I_{ca}. J Pharmacol Exp Ther 308:324–329

Sugawara A, Imamura T, Aso S, Ebitani K (1997) Change of paralytic shellfish poison by the marine bacteria living in the intestine of Japanese surf clam, *Pseudocardium sybillae* and the brown sole, *Pleuronectes herensteini*. Sci Rep Hokkaido Fisher Exp Stat 50:35–42

Sullivan JJ (1982) Paralytic shellfish poisoning: analytical and biochemical investigation. PhD thesis, University of Washington, Seattle.

Taylor FJR (1984) Toxic dinoflagellates: taxonomic and biogeographic aspects with emphasis on *Protogonyaulax*. In: Ragelis EP (ed) Seafood toxins, vol 262, Amer. Chem. Soc. Symposium Ser. ACS, Washington, DC, pp 77–97

Teste V, Briand JF, Nicholson BC, Puiseux-Dao S (2002) Comparison of changes in toxicity during growth of Anabaena circinalis (cyanobacteria) determined by mouse neuroblastoma bioassay and HPLC. J Appl Phycol 14:399–407

Usleber E, Dietrich R, Burk C, Schneider E, Martlbauer E (2001) Immunoassay methods for paralytic shellfish poisoning toxins. J Assoc Off Anal Chem 84(5):1649–1656

Usup G, Kulis DM, Anderson DM (1994) Growth and toxin production of the toxic dinoflagellate Pyrodinium bahamense var. compressum in laboratory cultures. Nat Toxins 2:254–262

Usup G, Kulis DV, Anderson DM (1995) Toxin production in a Malaysian isolate of the toxic dinoflagellate *Pyrodinium bahamense var. compressum*. In: Lassus P, Arzul G, Erard E, Gentien P, Marcailiou C (eds) Harmful marine algal blooms. Lavoisier, Paris, pp 519–524

Usup G, Pin LC, Ahmad A, Teen LP (2002) Alexandrium (Dinophyceae) species in Malaysian waters. Harmful Algae 1:265–275

Vaulot D, Lebot N, Marie D, Fukai E (1996) Effect of phosphorus on Synechococcus cell cycle in surface Mediterranean waters during summer. Appl Environ Microbiol 62:2527–2533

Velzeboer RMA, Baker PD, Rositano J, Heresztyn T, Codd GA and Raggett SL (2000) Geographical patterns of occurrence and composition of saxitoxins in the cyanobacterial genus Anabaena (*Nostocales, Cyanophyta*) in Australia. Phycologia 39:395–407

Vieytes MR, Cabado AG, Alfonso A, Louzao MC, Botana AM, Botana LM (1993) Solid phase radioreceptor assay for paralytic shellfish toxins. Anal Biochem 211:87–93

Wang JX, Salata JJ, Bennetn PB (2003) Saxitoxin is a gating modifier of hERG K+channels. J Gen Physiol 121:583–598

Wang S, Tang D, He F, Fukuyo Y, Azanza RV (2008) Occurrences of harmful algal blooms (HABs) associated with ocean environments in the South China Sea. Hydrobiology 596:79–93

Wiese M, D'Agostino PM, Mihali TK, Moffitt MC, Neilan BA (2010) Neurotoxic alkaloids: saxitoxin and its analogs. Mar Drugs 8:2185–2211

World Health Organization (WHO) (1984) Environmental Health Criteria 37: Aquatic (Marine and Freshwater) Biotoxins.

Xie W, Liu X, Yang X, Zhang C, Bian Z (2013) Accumulation and depuration of paralytic shellfish poisoning toxins in the oyster *Ostrea rivularis* Gould - chitosan facilitates the toxin depuration. Food Control 30:446–452

Yoshida T, Sako Y, Kakutani T, Fujii A, Uchida A, Ishida Y, Arakawa O, Noguchi T (1998) Comparative study of two sulfotransferases for sulfation to N-21 of *Gymnodinium catenatum* and *Alexandrium catenella* toxins. In: Reguera B, Blanco J, Fernández ML, Wyatt T (eds) Harmful algae. Xunta de Galicia and IOC, United Nations Educational, Scientific, and Cultural Organization, Grafisant, Spain, pp 366–369

Sources, Fluxes, and Biogeochemical Cycling of Silver in the Oceans

Céline Gallon and A. Russell Flegal

Contents

1 Introduction 27
2 Oceanic Silver: Data and Data Gaps 29
3 Measurements of Silver in the Oceans 33
    3.1 Waters 33
    3.2 Sediments 39
    3.3 Qualifications 42
4 Summary 43
References 44

1 Introduction

This brief review of silver in the oceans was catalyzed by our observation that there was a relative paucity of published data on the topic. A few authors have reported silver concentrations in the North Pacific (Martin et al. 1983; Zhang et al. 2001, 2004; Ranville and Flegal 2005; Kramer et al. 2011), South Pacific (Murozumi 1981; Zhang et al. 2004), North Atlantic (Flegal et al. 1995; Rivera-Duarte et al. 1999; Ndung'u et al. 2001), South Atlantic (Flegal et al. 1995; Ndung'u et al. 2001), and Southern Oceans (Miller and Bruland 1995; Sañudo-Wilhelmy et al. 2002), as well as in the Bering (Zhang et al. 2004) and Baltic (Ndung'u 2011) Seas. We are not aware, however, of any published data on silver concentrations in the Indian Ocean, and there are only a few reports on silver concentrations in marine sediments.

C. Gallon (✉) • A.R. Flegal
Institute of Marine Sciences, University of California, Santa Cruz, CA 95064, USA
e-mail: celine_gallon@yahoo.com; flegal@ucsc.edu

That scarcity of data on silver concentrations in the oceans raises the question: Who cares? Silver was not included in the original list of elements for GEOTRACES, "an international programme which aims to improve the understanding of biogeochemical cycles and large-scale distribution of trace elements and their isotopes in the marine environment" (GEOTRACES 2006), because it is neither a nutrient nor a commonly used tracer in sea water. However, silver is extremely toxic to marine phytoplankton and invertebrates, and its increasing use as a biocide has raised concerns about its potential as an environmental pollutant (Purcell and Peters 1998).

Silver has been extracted from geological deposits since 300 BC (Nriagu 1996). It has been mined directly, as well as extracted as a bi-product of gold, copper, lead and zinc deposits. Ancient Greeks and Romans extracted considerable amounts of silver for the production of silver bullion, and it has been used widely throughout history for the production of currencies, ornaments and utensils. Technological innovations since the 1800s have led to new uses of silver. In particular, its photochemical properties were the basis for the development of photography.

However, environmental releases of industrial silver from industrial and municipal wastewater outfalls, and from mining, smelting, and manufacturing operations have adversely impacted the environment (Purcell and Peters 1998), primarily because of silver's toxicity to invertebrates (Luoma et al. 1995). For example, wastewater discharges of silver from a photography plant into San Francisco Bay measurably decreased the fecundity of benthic invertebrates in the effluent plume (Flegal et al. 2007). Although such discharges have substantially decreased in the United States since the 1970s, industrial silver emissions are likely to have increased in rapidly developing countries that possess limited silver recovery and treatment facilities (Ranville and Flegal 2005).

Moreover, concerns with silver toxicity in aquatic environments have markedly increased with the rapidly growing use of silver nanoparticles (AgNPs) as antimicrobial agents in a wide variety of consumer products (Blaser et al. 2008; Luoma 2008; Bradford et al. 2009; Fabrega et al. 2011). As a result, there have recently been a series of studies on AgNP's stability (e.g., Benn and Westerhoff 2008; Liu and Hurt 2010; Liu et al. 2010; Levard et al. 2012; Unrine et al. 2012; He et al. 2013; Chambers et al. 2014), bioavailability to aquatic organisms (e.g., Li et al. 2013; Wang and Wang 2014), and toxicity to those organisms (e.g., Navarro et al. 2008; Bone et al. 2012; Turner et al. 2012; He et al. 2013; Chambers et al. 2014). These studies build on previous reports that addressed the bioaccumulation of silver in marine food chains (e.g., Ettajani et al. 1992; Fisher et al. 1995; Fisher and Wang 1998; Xu and Wang 2004; Yoo et al. 2004; Long and Wang 2005; Ng and Wang 2007). Those and other reports are summarized in Eisler's (2010) recent compendium on silver concentrations in marine organisms. Ratte (1999) and Bianchini et al. (2005) also provided earlier reviews on the bioaccumulation and toxicity of silver in marine organisms, and there have been several complementary reports since then (e.g., Bianchini et al. 2005; Pedroso et al. 2007).

Silver, in its ionic form, is extremely toxic to some aquatic organisms, second to only mercury (Luoma et al. 1995). As such it is listed by the U.S. Environmental Protection Agency (USEPA) as a priority pollutant in natural waters. But, in marine waters, most silver is associated with chloro-complexes and very little free ion is present (Miller and Bruland 1995). Notwithstanding, silver is rapidly bioaccumulated from solution by some marine invertebrates to potentially toxic levels (Engel et al. 1981; Luoma et al. 1995; Wang 2001).

The nature of silver's biogeochemical cycle in the oceans is still poorly understood. Its vertical concentration profiles in the oceans exhibit typical nutrient-like behavior with similarities to that of silicate, suggesting that their biogeochemical cycles are linked (Martin et al. 1983; Flegal et al. 1995; Ndung'u et al. 2001). However, there are pronounced deviations between the profiles of silver and silicate, which renders them less analogous (Zhang et al. 2004; Ranville and Flegal 2005; Kramer et al. 2011). Moreover, recent studies indicate that the anthropogenic inputs of silver to the oceans may be relatively substantial, compared to natural inputs, and in contrast to those of silicate, which are dominated by natural sources. Consequently, it has been proposed that spatial and temporal gradients of silver concentrations in marine aerosols and surface waters may be used to distinguish between natural and industrial inputs of silver and associated industrial contaminants (e.g., selenium) to the oceans (Ranville et al. 2010).

2 Oceanic Silver: Data and Data Gaps

Although data on the content in seawater of most trace metals (e.g., Cd, Cu, Fe, Hg, Ni, Pb, Zn) are limited (GEOTRACES 2006), there are even fewer measurements of silver in the open ocean (see Table 1 and Fig. 1). As previously noted, we found only a handful of peer-reviewed reports of silver in the oceans, and no reports of silver in the Indian Ocean. Similarly, we found very few peer-reviewed reports on silver in marine sediments (Table 2).

Some data on silver in oceanic hydrothermal plumes have been published, including one in the Mid-Atlantic Ridge (Douville et al. 2002). Concentrations of silver, copper and silicate found in those plumes are summarized in Table 3. This table shows anomalously higher concentrations of silver in sulfidic waters (4–51 nmol/kg) than in other oceanic waters (0.2–87.7 pmol/kg). From this disparity, we believe hydrothermal inputs may be relatively important sources in the budget and cycling of silver within the oceans.

In contrast to the paucity of data for dissolved silver concentrations in oceanic waters, there is a relative wealth of data on silver in estuarine waters—at least in San Francisco Bay (Sañudo-Wilhelmy et al. 2004). There, silver concentrations in total dissolved (<0.45 m) and total (unfiltered) surface waters and sediments have been systematically measured for more than three decades (Flegal and Sañudo-Wilhelmy 1993; Smith and Flegal 1993; Sañudo-Wilhelmy et al. 1996; Rivera-Duarte and Flegal 1997; Spinelli et al. 2002; Squire et al. 2002; Flegal et al. 2007; Huerta-Diaz et al. 2007). In addition, a few other measures of silver in other estuarine and coastal waters exist (Sañudo-Wilhelmy and Flegal 1992; Wen et al. 1997; Buck et al. 2005; Clark et al. 2006; Beck and Sañudo-Wilhelmy 2007; Cozic et al. 2008; Godfrey et al. 2008; Zhang et al. 2008; Tappin et al. 2010) (Table 4).

Table 1 Reported range of silver concentrations (pmol/kg) in oceanic waters

Sample site	Sampling date	[Ag] range	Fraction	Filter size	UV	Chemistry	Analytical method	Reference
Baltic Sea	2005–2008	<1–9.4 (surface)	Filtered	0.4 μm	Yes	Chelating resin extraction	ICP-MS	Ndung'u (2011)
North Atlantic	1990	0.3–7.2	Total	–	No	Solvent extraction	GFAAS	Flegal et al. (1995)
	1993	0.7–6.9	Total	–	No	Solvent extraction	GFAAS	Rivera-Duarte et al. (1999)
	1996	1.7–10.7	Filtered	0.22 μm	No	Solvent extraction	ICP-MS	Ndung'u et al. (2001)
South Atlantic	1990	0.2–9.6	Total	–	No	Solvent extraction	GFAAS	Flegal et al. (1995)
	1996	1.2–31.7	Filtered	0.22 μm	No	Solvent extraction	ICP-MS	Ndung'u et al. (2001)
Southern Ocean	1991	8.9–22.4 (surface)	Filtered	0.45 μm	No	Solvent extraction	GFAAS	Sañudo-Wilhelmy et al. (2002)
	1992	5–8 (30 m)	Filtered	0.2 μm	No	Solvent extraction	GFAAS	Miller and Bruland (1995)
Indian Ocean	–	–	–	–	–	–	–	–
South Pacific	–	1.0–32.0	Filtered	–	No	Solvent extraction	ID-TIMS	Murozumi (1981)
	2001	9–11.4	Filtered/total	0.2 μm	No	Solvent extraction	ID-ICPMS	Zhang et al. (2004)
North Pacific	1981	0.4–25.0	Filtered	0.4 μm	No	Solvent extraction	GFAAS	Martin et al. (1983)
	1998	4.2–46.8	Filtered	0.45–0.04 μm	No	Solvent extraction	ID-ICPMS	Zhang et al. (2001)
	2000–2001	1.5–66.2	Filtered	0.04–0.2 μm	No	Solvent extraction	ID-ICPMS	Zhang et al. (2004)
	2002	1.0–87.7	Total	–	Yes	Chelating resin extraction	ICP-MS	Ranville and Flegal (2005)
	2009	6.2–72.9	Total	–	Yes	Chelating resin extraction	ICP-MS	GEOTRACES IC cruise
	2005	6.0–71.6	Total/filtered	–	No	Solvent extraction	ID-ICPMS	Kramer et al. (2011)
Bering Sea	1999	2.7–104.5	Filtered	0.04 μm	No	Solvent extraction	ID-ICPMS	Zhang et al. (2004)

GFAAS graphite furnace atomic absorption spectrometry, *ICP-MS* inductively coupled plasma—mass spectrometry, *ID-TIMS* isotope dilution—thermal ionization mass spectrometry, *ID-ICPMS* isotope dilution—inductively coupled plasma—mass spectrometry

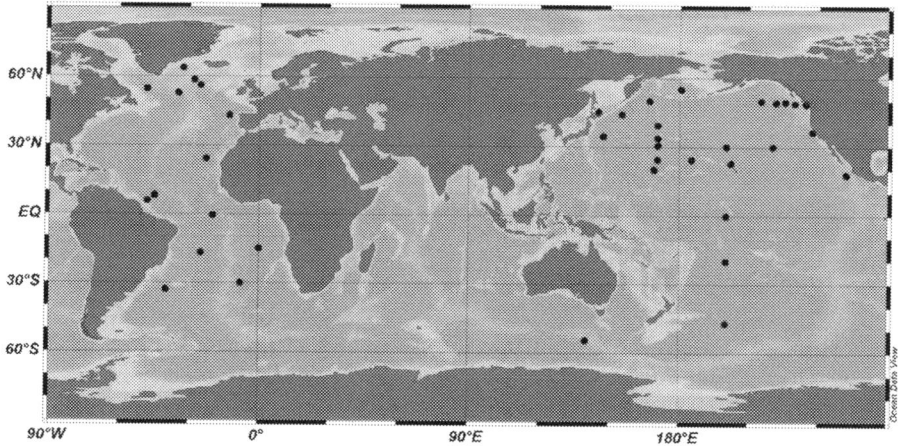

Fig. 1 Locations of reported depths profiles of silver concentrations

Table 2 Silver Concentrations (nmol/kg) in marine sediments

Location	[Ag] range		Reference
	Solid phase	Porewaters	
Coastal zones across the globe	12–2,201		Koide et al. (1986)
Puget Sound, Washington, USA	140–6,560		Bloom and Crecelius (1987)
Massachusetts and Cape Cod bays, Massachusetts, USA	280–8,250		Ravizza and Bothner (1996)
Madeira Abyssal Plain, eastern Mediterranean basin, northeast Atlantic[a]	~0–140		Crusius and Thomson (2003)
Namibian diatom belt	19 ± 10		Borchers et al. (2005)
Peru OMZ[b]	3–1,305		Böning et al. (2004)
Chile OMZ[b]	8–46		Böning et al. (2005, 2009)
Coastal Massachusetts	6,000–24,000		Kalnejais et al. (2007)
Western Canadian, Mexican, Peruvian and Chilean continental margins	740–13,750		McKay and Pedersen (2008)
Washington/Oregon states, USA	190–6,300	0.03–25	Morford et al. (2008)
Boston Harbor, Massachusetts, USA		0.03–0.17	Kalnejais et al. (2010)

[a]Extrapolated from graph
[b]Oxygen Minimum Zone

Those estuarine data documented that marked anthropogenic perturbations of the biogeochemical cycling of silver occurred in estuarine and neritic waters from surface runoff and point source discharges. The source of much of that discharged silver occurred after use in X-ray and photographic processing, solders, electronics, and as a bactericide and algaecide. Most of that silver is rapidly scavenged and deposited in benthic sediments (Benoit et al. 2010), but it may subsequently be

Table 3 Reported range of silver (nmol/kg), copper (μmol/kg), and silicate (μmol/kg) concentrations in Mid-Atlantic Ridge hydrothermal plumes (from Douville et al. 2002)

System name	Location	Collection date	[Ag]	[Cu]	[Si]
Menez Gwen	37°50′N	1994	4.3	<2	8.2
		1997	17	<2	11.2
Lucky Strike	37°17′N	1994, 1997	4.7–25	<2–30	8.2–16
Rainbow	36°14′N	1997	47	140	6.9
Broken Spur[a]	29°N	1993	–	43	–
TAG	23°N	1996	51	130	20
Snakepit	23°N	1996	31	35	20
Logatev	34°45′N	1996	11	27	8.2

[a]From James et al. (1995)

Table 4 Dissolved silver concentrations (pmol/kg) in coastal and estuarine waters

Location		[Ag] Range	Reference
North America			
	San Francisco Bay	<1–244	Squire et al. (2002)
	San Diego Bay	66–307	Sañudo-Wilhelmy and Flegal (1992)
	Hudson River	<–102	Godfrey et al. (2008)
	Long Island Sound	3–353	Buck et al. (2005)
	Great South Bay	<1–73	Clark et al. (2006)
	Jamaica Bay	2–452	Beck and Sañudo-Wilhelmy (2007)
	Trinity River estuary	4–59	Wen et al. (1997)
Europe			
	Restronguet Creek	25–190	Barriada et al. (2007)
	Adriatic Sea	14–28	"
	Seine Estuary	ND–60	Cozic et al. (2008)
	Gulma Fjord	53–12	Tappin et al. (2010)
	Adriatic Sea	12–33	"
	Tamar Estuary	ND–17	"
	Fal Estuary	14–34	"
	Restronguet Creek	32–181	"
	River Mero Estuary	2–40	"
	A Coruña Bay	23–115	"
Asia			
	Tokyo Bay	Jun–15	Zhang et al. (2008)

ND = Non Detect

remobilized by natural digenetic and/anthropogenic processes, which also recycle legacy inputs of industrial silver from acid mine drainage and other historic inputs (Smith and Flegal 1993; Rivera-Duarte and Flegal 1997; Tappin et al. 2010). Fortunately, the advent of digital photography has markedly decreased discharges of silver from photographic processing, and that has resulted in recent reductions of industrial silver inputs to the environment (Squire et al. 2002; Flegal et al. 2007).

But as previously noted, environmental inputs of silver nanoparticles have been increasing exponentially, and the bioavailability and toxicity of those materials are poorly understood.

Silver in estuarine waters generally exhibits a non-conservative behavior, with dissolved silver concentrations decreasing with salinity (Smith and Flegal 1993; Wen et al. 1997; Zhang et al. 2008). This pattern has been attributed to silver's high affinity for suspended particulates, as demonstrated by its relatively high partition coefficient (Kd ~10^5) (Sañudo-Wilhelmy et al. 1996; Wen et al. 1997; Zhang et al. 2008; Tappin et al. 2010). At low salinities, silver is strongly associated with iron and manganese oxyhydroxide/sulfide phases, organic macromolecules, and colloids (Wen et al. 1997; Reinfelder and Chang 1999). In more saline waters, the macromolecular fraction decreases and dissolved silver chloro-complexes become more important (Turner et al. 1981; Miller and Bruland 1995). However, some "dissolved" (<0.45 μm) silver remains tightly bound to refractory organics in estuarine and marine waters (Ndung'u et al. 2006).

Silver may also be present as a soluble sulfide in fresh and estuarine waters, as reported by Rozan and Luther (2009). They proposed this was due to the high association constant of silver sulfides (pK 12-30) for multi-nuclear metal clusters with stoichiometries of 2:1 and 3:3. The presence of those silver sulfides was associated with anoxic (sulfidic) sediments and wastewater discharges, which would not be a factor in most oceanic waters. However, those complexes could be important in marine hydrothermal plumes.

3 Measurements of Silver in the Oceans

3.1 Waters

Silver concentrations observed in oceanic waters display marked spatial differences (Fig. 2). Recorded silver concentrations show a systematic increase along the oceanic conveyor belt circulation, whereas the nutrient-type vertical profile of silver is retained (Bruland and Lohan 2004). As previously noted, silver's nutrient-type distribution is evidenced by its covariance with the distribution of silicate, which is illustrated in Fig. 3, suggesting that both geochemical cycles are linked (Martin et al. 1983; Flegal et al. 1995; Ndung'u et al. 2001).

Silver is hypothesized to be sequestered within a refractory organic phase that is associated with biogenic silica, and to follow a parallel biological scavenging and subsequent remineralization (Ndung'u et al. 2006). However, departures from the linear correlation between the two elements emphasize that the exact mechanisms of the silver marine cycle are still poorly understood (Zhang et al. 2004; Ranville and Flegal 2005; Kramer et al. 2011). Silver levels exceed those of silica when silver concentrations are ≥20 pmol/kg, suggesting that silver may be relatively enriched from aeolian deposition and/or is more slowly remineralized than silica. There may also be a relatively greater diagenetic remobilization of silver from bottom sediments,

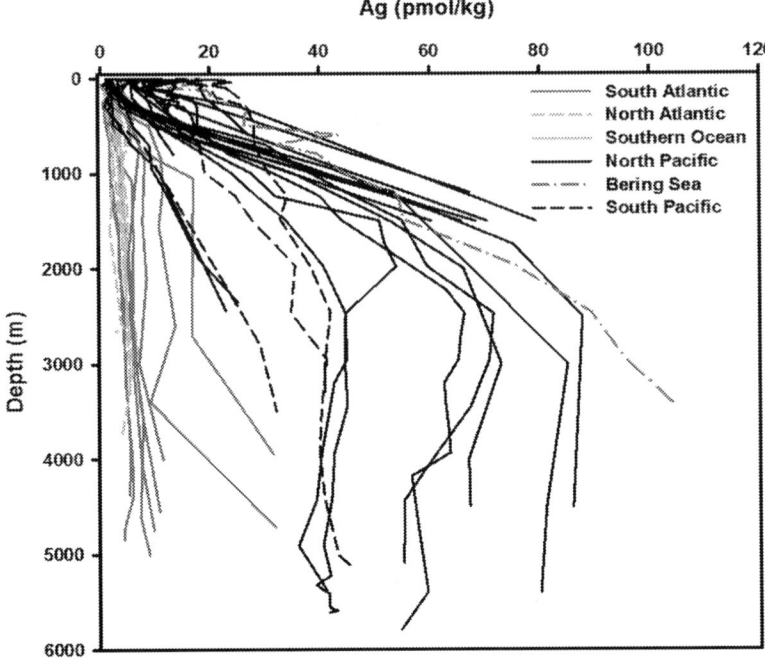

Fig. 2 Vertical profiles of silver concentrations (pmol/kg) in the oceans. Data from Murozumi (1981), Martin et al. (1983), Flegal et al. (1995), Rivera-Duarte et al. (1999), Ndung'u et al. (2001), Sañudo-Wilhelmy et al. (2002), Zhang et al. (2001, 2004), Ranville and Flegal (2005), and Kramer et al. (2011). Also included are previously unpublished data from our laboratory

a more efficient transfer of silver out of surface waters in sinking particle aggregates, and/or a relatively greater input (e.g., hydrothermal vents) of silver in deep oceanic waters.

In addition to silica, silver concentrations in seawater profiles also correlate with those of copper (Fig. 4). The position of silver immediately under copper in the Periodic Table attests to its role as a biogeochemical analog of copper, an essential trace element. This similarity also accounts for some of silver's toxicity to many marine invertebrates, whose respiratory pigment is the copper-based hemocyanin (Burmester 2002). However, as with silica, silver concentrations deviate from a true linear correlation with those of copper, especially at higher concentrations—again attesting to dissimilarities in the biogeochemical cycles of the two elements.

Those differences are illustrated in Table 3, which lists concentrations of silver, silicate and copper that are present in hydrothermal plumes. Based on the sea water concentrations of silver (0.023 nmol/kg) and copper (0.0033 µmol/kg), as given in the original compilation by Douville et al. (2002), the atomic ratio of silver to copper in the plumes ($3.4–8.9 \times 10^{-4}$) is one order of magnitude lower than it is in sea water. Consequently, the relative excess of silver compared to copper in deep ocean water is not due to a simple enrichment from hydrothermal inputs. But differences

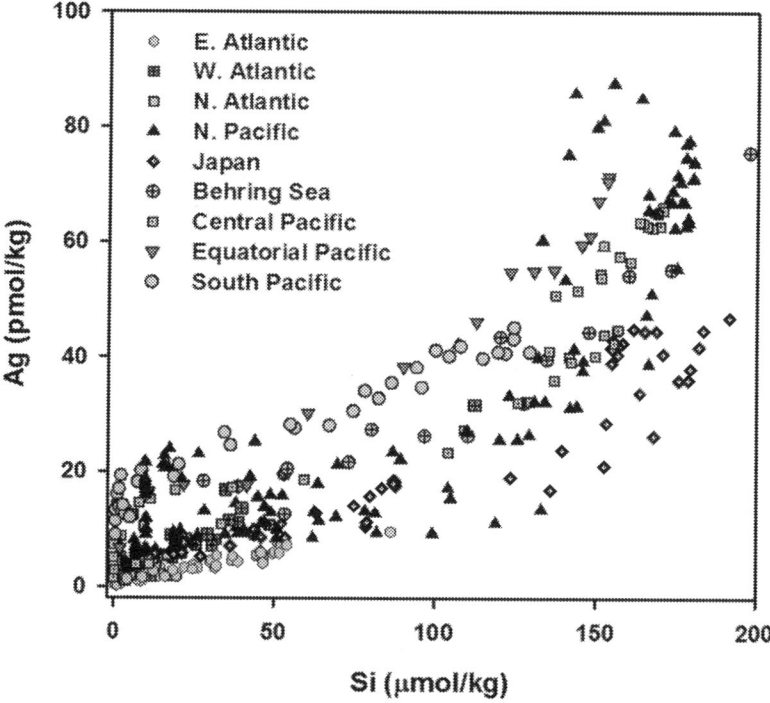

Fig. 3 Plot of silver (pmol/kg) vs. silicate (μmol/kg) in oceanic waters. Data from Flegal et al. (1995), Rivera-Duarte et al. (1999), Ndung'u et al. (2001), Zhang et al. (2001, 2004), Ranville and Flegal (2005), and Kramer et al. (2011)

in the solubility, speciation, and dispersion of silver and copper could subsequently enrich silver, compared to copper in those plumes, and that enrichment could extend well beyond the hydrothermal sources, as determined for other trace elements (e.g., manganese).

The most recent silver data in seawater have been mostly derived from the North Pacific Ocean, and have prompted alternate and contrasting hypotheses on the origins and cycling of silver in oceanic waters. Figure 5 synthesizes all of the reported silver vertical concentration profiles in the North Pacific (Murozumi 1981; Martin et al. 1983; Zhang et al. 2001, 2004; Ranville and Flegal 2005; Kramer et al. 2011), excluding those from northeast Pacific coastal and estuarine waters (Bloom and Crecelius 1984; Flegal and Sañudo-Wilhelmy 1993; Smith and Flegal 1993; Rivera-Duarte and Flegal 1997; Squire et al. 2002; Flegal et al. 2007). These data show that recent silver concentrations in water samples collected at intermediate depth (2,400–2,500 m) (Zhang et al. 2001, 2004; Ranville and Flegal 2005; Kramer et al. 2011) are up to four-fold greater than those previously measured in eastern North Pacific waters (Martin et al. 1983). The disparity may, in part, result from differences in sampling protocols (e.g., filtered versus unfiltered samples) and/or analytical protocols (e.g., UV versus non UV), as mentioned below.

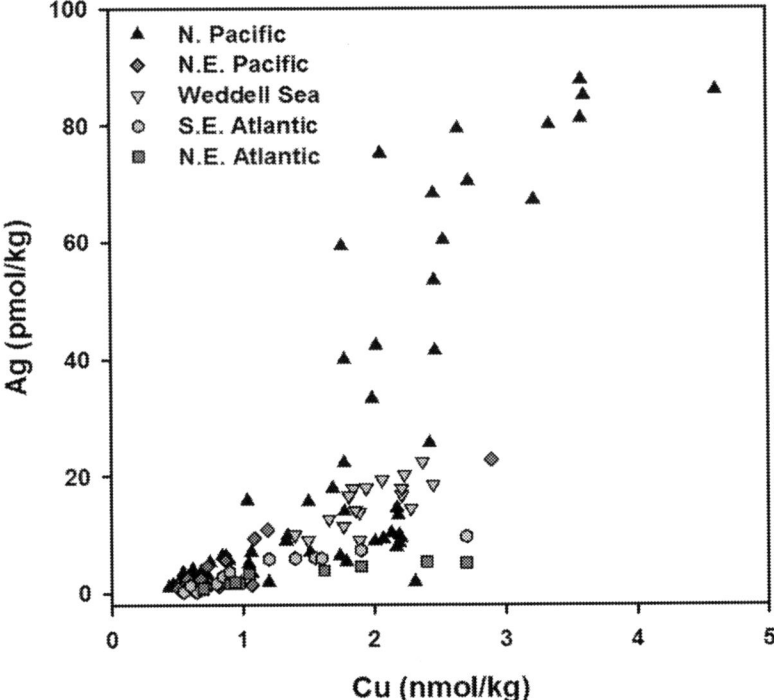

Fig. 4 Plot of silver (pmol/kg) vs. dissolved copper (nmol/kg) in oceanic waters. Data from Martin et al. (1983), Flegal et al. (1995), Sañudo-Wilhelmy et al. (2002), and Ranville and Flegal (2005)

Other factors have been advanced to explain the spatial variation of silver vertical concentration profiles in the North Pacific and the observed departures from the linear correlation between silver and silica. Ranville and Flegal (2005) proposed that the discrepancies observed between measurements in waters collected in 2001 in the western North Pacific and those previously measured in the eastern North Pacific could be explained by the incorporation into intermediate waters in the Sea of Okhotsk of surface waters enriched by atmospheric inputs of contaminant silver aerosols. Silver concentrations in surface waters collected near the Asian mainland during the same cruise had values as high as 12 pmol/kg, higher than any previously reported value for the open ocean (Fig. 5). Concentrations then steadily decreased eastward to levels of 1–2 pmol/kg near the central part of the North Pacific Gyre (Fig. 6). Since this eastward decrease corresponds with the prevailing westerly wind flow from the Asian mainland, it was suggested that atmospheric transport of mineral and industrial aerosols over the North Pacific is responsible for elevated silver concentrations in these surface waters.

The enrichment of silver over aluminum and the lack of correlation between the two elements in surface waters suggested a dominance of atmospheric fluxes of industrial emissions rather than natural processes—with the qualification that the biogeochemical processes governing the residence times of silver and aluminum in oceanic

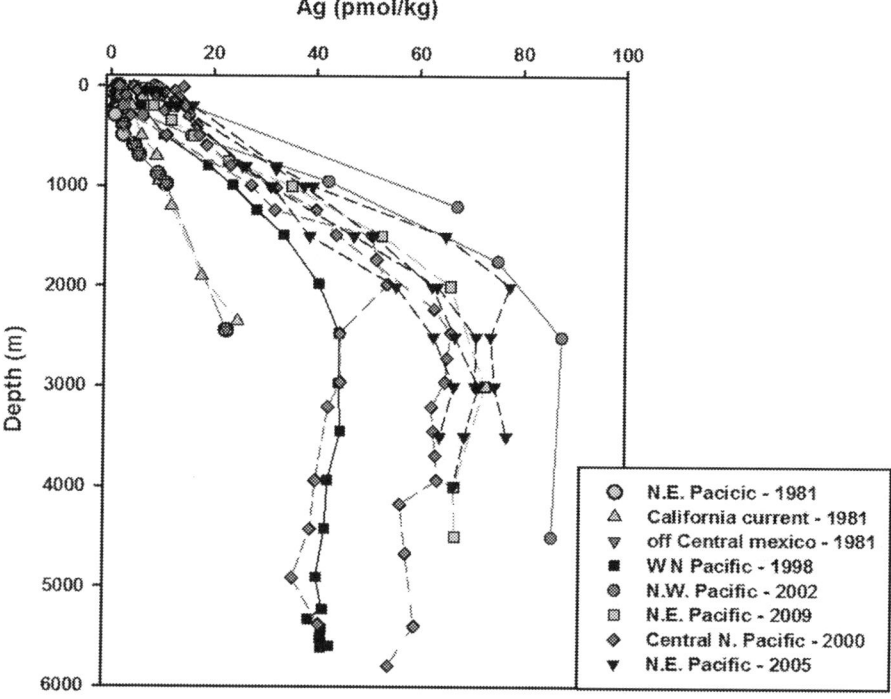

Fig. 5 Temporal variations of silver concentrations (pmol/kg) measured in waters collected in the North Pacific between 1981 and 2009. Data from Martin et al. (1983), Zhang et al. (2001, 2004), Ranville and Flegal (2005), and Kramer et al. (2011). Also included are previously unpublished data from our laboratory

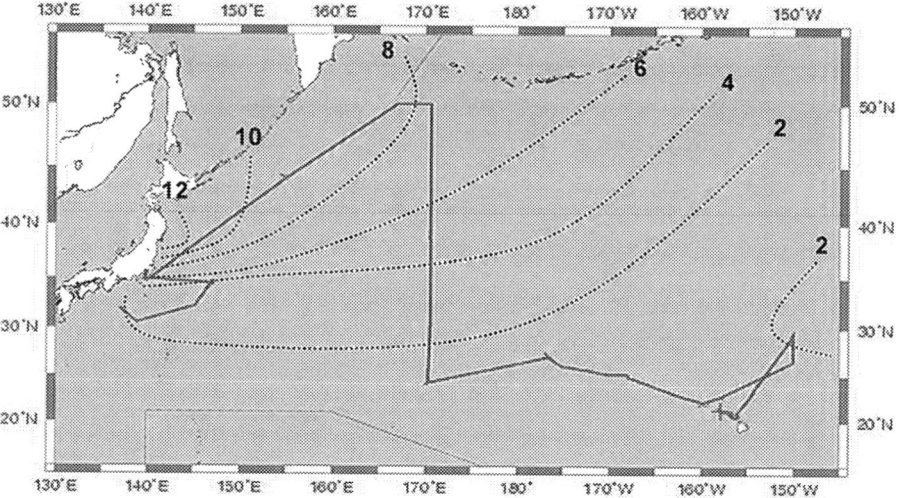

Fig. 6 Oceanic enrichment of silver. Contour plot of surface water silver concentrations (pmol/kg) in the North Pacific Ocean. The *solid line* indicates the ship's cruise track; contours are extrapolated from data collected along the cruise track (from Ranville and Flegal 2005)

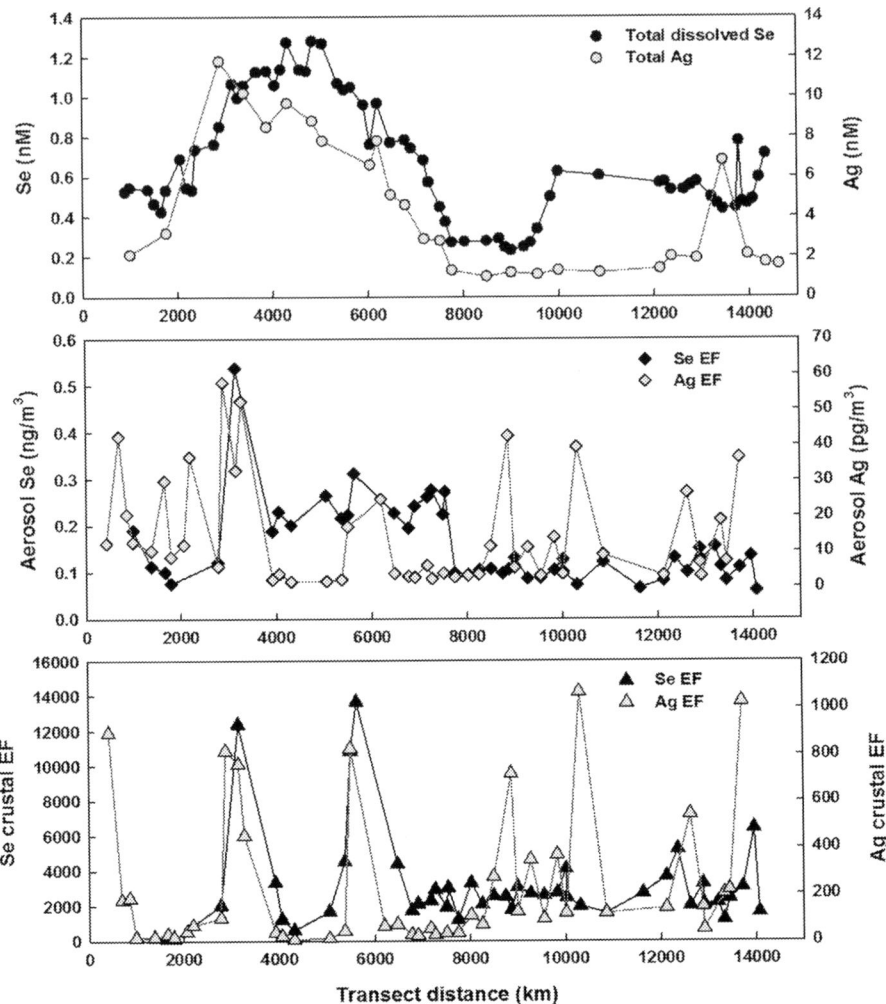

Fig. 7 Covariance of silver and selenium concentrations in surface waters and aerosols, along with their corresponding enrichment factors (EF), in the North Pacific Ocean (from Ranville et al. 2010)

surface waters are quite dissimilar. A similar gradient was also observed in lead concentrations and isotopic compositions in surface waters from the same cruise, and these were traced to Asian industrial emissions (Gallon et al. 2011). That spatial gradient of silver was comparable to the gradient observed in the North Atlantic by Rivera-Duarte et al. (1999), who also hypothesized that relatively elevated silver concentrations in those surface waters were from external inputs, although it was unclear whether this elevation results from atmospheric or terrestrial inputs.

Silver concentrations in North Pacific surface waters were also found to parallel those of selenium, which is both an essential trace element and a contaminant resulting from coal combustion (Ranville et al. 2010), as illustrated in Fig. 7. This similarity further suggested the predominance of aeolian fluxes of silver from Asian industrial emissions to the North Pacific—again with the qualification that the biogeochemical processes governing the residence times of silver and selenium in oceanic surface waters are also quite dissimilar. Because natural emissions measurably contribute to the aeolian fluxes of selenium—but are presumably negligible for silver—in the North Pacific, variations in the [Ag]:[Se] ratios may be used to distinguish between natural and industrial fluxes of selenium to those oceanic waters.

However, Zhang et al. (2004) proposed that different, natural processes accounted for the distribution of silver in the northwest Pacific. They observed that silver concentrations increased with latitude in the surface waters of the Bering Sea, the central North and South Pacific and the Southern Ocean. They attributed the increased silver in those waters to an upwelling and vertical mixing similar to that of silica. They also attributed elevated Ag/Si ratios in the surface vs. deep waters to differences in biological uptake. Then, they used a scavenging/regeneration model to characterize the Ag–Si relationship.

Finally, Kramer et al. (2011) recently proposed another natural process to account for the distribution of silver in the North Pacific. Their account was based on correlations between low oxygen and silver concentrations in the broad subsurface oxygen minimum zone (OMZ) of the North Pacific, which could produce the spatial variability observed in profiles in the central eastern Pacific. Specifically, they noted that Ag:Si profiles and oxygen profiles present relatively similar shapes, and that the intensity of the subsurface Ag/Si minima follows the same trend as that of the OMZ. In Fig. 8, we show Ag:Si and O_2 profiles in the Pacific Ocean that are available in the literature, as well as a plot of Ag:Si vs. O_2. These observations were interpreted by Kramer et al. (2011) as an indication that dissolved oxygen content could apply a secondary control on the dissolved silver concentration.

Kramer et al. (2011) also hypothesized that silver may be removed from oxygen-depleted waters by scavenging and/or precipitation of AgS species and subsequent sequestration in the underlying reducing sediments. Such mechanisms could be the result of an exchange of dissolved silver with thermodynamically less stable metal-sulphide nanoclusters (e.g., Cu, Cd, Zn) present in the oxic water column or its complexation with nanomolar concentrations of free sulphide, as described by Rozan and Luther (2009). In addition, silver may be scavenged in the water column and precipitated as Ag_2S in anoxic microenvironments within settling decaying organic matter, as proposed by McKay and Pedersen (2008).

3.2 Sediments

Similarly to waters, few researchers have investigated the distribution and biogeochemical cycling of silver in marine sediments. The relative paucity of information is illustrated by the fact that we found only ten peer-reviewed references on silver in marine sediments, which are listed in Table 2. This absence of information is surprising, because there is potential for silver (as a biogeochemical analogue of silicate) to be used as a proxy of past diatom fluxes to the sea floor. But the

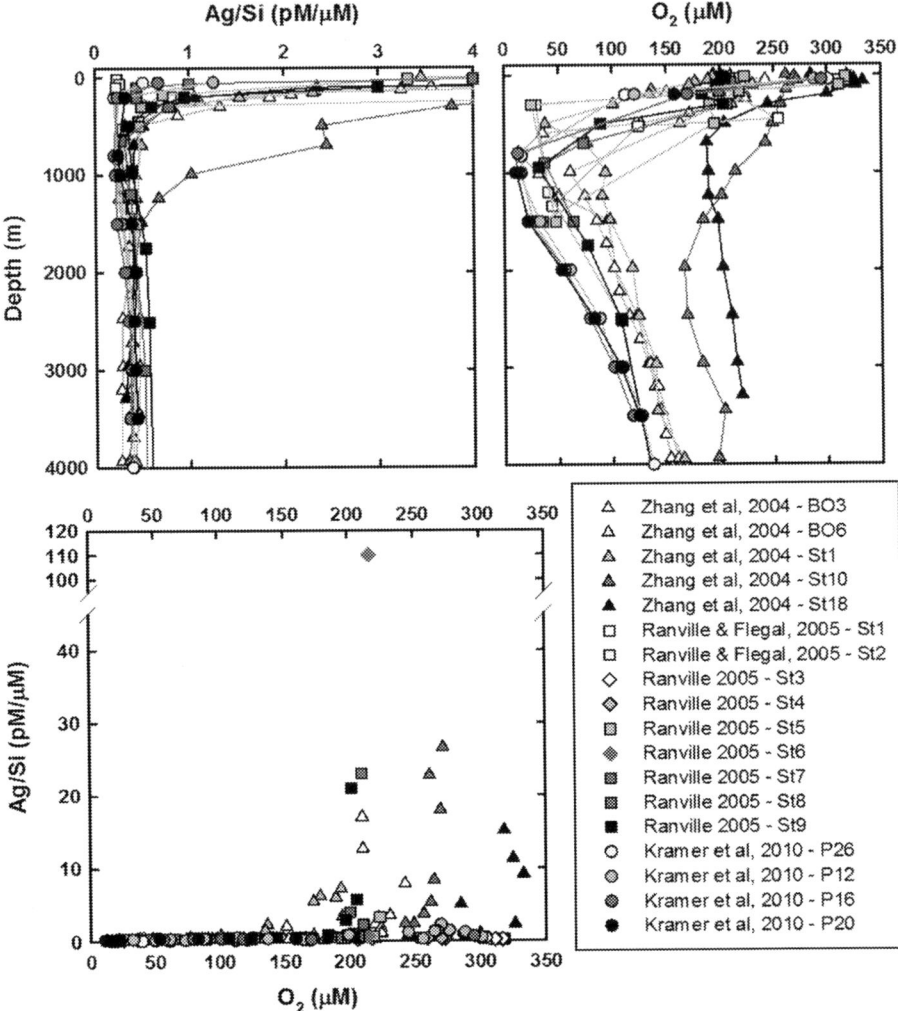

Fig. 8 Dissolved Ag:Si and O_2 profiles and correlations in the Pacific Ocean (Zhang et al. 2004; Ranville and Flegal 2005; Kramer et al. 2011)

deposition silver and its potential for biogenic remobilization in marine sediments is complex, as evidenced in San Francisco Bay (Smith and Flegal 1993; Rivera-Duarte and Flegal 1997).

That complexity is illustrated by studies of silver in marine sediments, which indicate that multiple factors and dynamic processes are involved. Early work by Koide et al. (1986) showed silver enrichment relative to its crustal abundance in coastal anoxic sediments collected around the globe. Later, Borchers et al. (2005) concluded that the enrichment they observed in Namibian upwelling sediments resulted from a biogenic pre-concentration prior to burial. Silver distributions in sediment cores collected within and around the oxygen minimum zone (OMZ) off

Peru could be related to the particle flux of diatomaceous matter to the sediment (Böning et al. 2004). In the suboxic sediment below the OMZ, the observation of 'low-opal sediments' and of a relationship between opal and silver concentrations was attributed to particle dissolution as they settle in the deep water column and the subsequent remineralization of opal and silver in equal proportions. 'High-opal sediments' were observed in anoxic sediments within and in the upper edge of the OMZ. Consequently, silver distributions there were likely related to an early diagenetic fixation with Total Organic Carbon (TOC) or Total Sulfur (TS) during opal dissolution, reflecting a redox-control of silver fixation.

Although organic matter and Ag_2S species seem to play a role in silver fixation, Crusius and Thomson (2003) suggested that it may also be sequestered by precipitation of silver selenide species (AgSe or Ag_2Se) as a result of oxygen exposure of sediments that were initially anoxic. In a subsequent study conducted off Chile, Böning et al. (2005) found that silver content in sediment increased with water depth, which they interpreted as an indication that silver enrichment was controlled by opaline regeneration and/or by a higher availability of silver in seawater. A redox control of silver content was discarded in that case, based on the contrast with Cd and Re contents, which decrease with increasing water depth and reflect decreasing reducing conditions.

The lack of redox influence on silver content was also observed by McKay and Pedersen (2008). They measured concentrations of silver in surface sediments from the Western Canadian, Mexican, Peruvian, and Chilean continental margins, and also observed that concentrations increased with water depth. The lack of correlation observed between the concentrations of silver and redox-sensitive trace metals (Re, Cd, and Mo) led them to conclude that silver accumulation was not controlled by sedimentary redox conditions. Instead, they hypothesized that silver accumulation results from its scavenging decaying organic particles as they settle through the water column. More specifically, they proposed that silver precipitates as Ag_2S within anoxic microenvironments that develop as a result of organic degradation inside sinking organic particles. In addition, McKay and Pedersen (2008) observed a positive correlation between silver and barium in surface and near surface sediments, suggesting similar mechanisms of enrichment. Because barium is commonly used as a tracer of paleoproductivity, these similarities suggest that silver may also be used as a paleoproxy.

Morford et al. (2008) further highlighted the highly dynamic process of silver diagenesis through the analysis of porewater and sediment profiles in the northeast Pacific. They observed background concentrations of silver in oxic sediments associated with high levels in pore waters, indicating a flux of silver from the sediments to the overlying water. In addition, increasing porewater concentrations with water column depth led them to theorize that silver is delivered to the sediments by scavenging—sorbing onto settling particles (a longer path leading to an associated higher content of silver), and subsequently being released to pore waters. Similarly, erosion chamber experiments by Kalnejais et al. (2010) showed that silver can be released from sediments to the dissolved phase during erosion events, and over longer time scales in association with the reaction of suspended particles in the water column.

3.3 Qualifications

A major problem in understanding the biogeochemistry of silver in the world oceans, apart from the paucity of data, is the difficulty in comparing the different data sets. The few measurements of truly trace concentrations of silver in the oceans have been conducted by different investigators using different sampling, treatment, and analytical methods (Table 1). For example, onboard collections have been conducted using different sampling systems, and have included—or not—a filtration of the samples on various types and sizes of filters. The disparities extend to the laboratory, wherein two types of methods have primarily been used to separate and pre-concentrate silver: solvent-extraction and chelating resin column partitioning (Table 1). One method, developed in the 1970s (Danielsson et al. 1978, Bruland et al. 1979), uses a combination of 1-pyrrolidinedithiocarbamate and diethyldithiocarbamate (APDC/DDDC) to chelate silver, followed by a double extraction into chloroform, and back-extraction into nitric acid. Another method, developed more recently (Ndung'u et al. 2006), consists of separating silver on a mini-column packed with a strong anion exchange resin (Dowex 1-X8) connected to a flow-injection system that allows on-line determination by high resolution inductively coupled mass spectrometry (HR ICP-MS). Different analytical methods have also been used for the analysis of silver concentrations, including thermal ionization mass spectrometry (TIMS), atomic absorption spectrometry (AAS), and HR ICP-MS.

In addition, the use—or not—of ultraviolet (UV) irradiation prior to the analysis may introduce some disparity among silver measurements. Ndung'u et al. (2006) showed that a digestion with UV radiation is necessary prior to silver analysis with anion exchange resin to liberate any metal bound in refractory organic complexes. Analysis of waters from the San Francisco Bay estuary and the North Pacific Ocean showed a 10–70% increase in silver concentrations measured after exposure to UV radiation. Similarly, comparable measurements on the Safe D2 and GEOTRACES GDI intercalibration samples showed an increase of silver following UV irradiation (unpublished data). Whether such discrepancies apply to the use of APDC/DDDC extraction is not clear.

Comparison of the silver data generated by different laboratories is further complicated by the lack of a certified value for silver in any reference seawater material. The National Research Council of Canada (NRC-CNRC) certified reference material for trace metals in estuarine water (SLEW-3) has an "information" value for silver, but it is not certified. To date, only two laboratories (including ours) have reported silver concentrations in the GEOTRACES North Pacific (SAFe Reference Samples) and North Atlantic (GEOTRACES Reference Samples) intercalibration waters, and these measurements have not been published (Ken Bruland, personal communication). But a few authors have reported numbers for the NRC-CNRC reference materials for trace elements in nearshore seawater (CASS-4) and estuarine water (SLEW-3), which are listed in Table 5.

Finally, there have been a few measurements of silver isotopes in the marine environment, but they have not been substantiated by intercalibration. Folsom et al.

Table 5 Silver concentrations (nmol/kg) reported in National Research Council Canada (NRC-CNRC) reference material for trace metals in nearshore seawater (CASS-4) and estuarine water (SLEW-3)

Reference	Ag (pmol/kg)	
	CASS-4	SLEW-3
Ndung'u et al. (2001)	56.6 ± 8.3	
Yang and Sturgeon (2002)	50.2 ± 0.7	17.9 ± 0.8
Ndung'u et al. (2006)	78.6 ± 18.2	27.5 ± 7.1
Ranville and Flegal (2005)	101 ± 2.6	36.3 ± 0.3
Morford et al. (2008)	53 ± 3; 54 ± 6; 59 ± 10	
Ndung'u (2011)		22.5 ± 1.5

(1970) used ratios of radiosilver (^{110m}Ag/^{108m}Ag) associated with the 1961–1962 atmospheric nuclear weapons testing period to date marine organisms; then, two of these authors (Hodge and Folsom 1972) used those ratios and ^{90}Sr/^{110m}Ag ratios in particulate aerosols to independently estimate the global budget of atmospheric "fallout" ^{110m}Ag from that testing period. Murozumi et al. (1981) measured silver concentrations, but not stable silver isotopic compositions in seawater by using isotope dilution TIMS. Recent advances, however, have demonstrated that silver isotopic ratios may be accurately and precisely measured in environmental materials (Schönbächler et al. 2007; Luo et al. 2010). Therefore, we expect that silver isotopic measurements will soon be employed to trace natural and industrial fluxes of silver and its biogeochemical cycling within the oceans.

4 Summary

Despite its relatively high acute toxicity to marine phytoplankton and invertebrates and its increasing use as a biocide, silver measurements in seawaters are still few and far apart. That paucity of data limits our understanding of its global biogeochemical cycle and an assessment of its potential applicability as a tracer of anthropogenic contamination of the oceans. The need for such an assessment has been heightened by the dramatic increase in AgNPs over the past decade.

Available measurements indicate that silver has a nutrient-type behavior in the oceans, and preliminary data indicate that its cycling is being perturbed by aeolian inputs of industrial silver. However the processes governing silver's biogeochemical cycle in the ocean are still subject to debate. For instance, precipitation/scavenging of AgS species in oxygen-depleted waters has been proposed to play a role in silver removal from ocean waters and subsequent accumulation in sediments. This hypothesis could be tested by undertaking additional measurements in waters and sediments in regions impacted by extensive oxygen minimum zones (OMZ). Furthermore, analysis of silver concentrations in waters impacted by hydrothermal plumes could resolve what their potential impact is on silver distribution in deep water masses.

In addition to the relative scarcity of reported silver concentration measurements in the oceans, comparing those few measurements that do exist is hampered by inadequate application of quality assurance and quality control steps during the original analyses. Future studies need to be calibrated with standard reference materials that have certified values for silver, and intercalibrated with measurements made by other institutions. Those criteria are benchmarks for the international GEOTRACES program, and they certainly apply to future new measurements of silver to be made in the oceans.

Finally, new tools are needed to resolve open questions about silver fluxes and biogeochemical cycling in the oceans. Variations in silver isotopic ratios ($\delta^{107/109}$Ag) have been reported between standard materials of different environmental media, including sediment, industrial and domestic sludge, and fish liver (Luo et al. 2010). Consequently, variations in that ratio may be used to fingerprint different sources of silver in the environment and study its biogeochemical cycle in the oceans. Similarly, more sophisticated models should be developed and used to assess the validity of the assumed chemical properties and behavior of silver compounds in the water column and sediments, and to improve our understanding of anthropogenic perturbations of its cycle and identify areas for additional measurements. Lastly, we endorse the suggestion by McKay and Pedersen (2008) that the applicability of sedimentary silver as a proxy of paleoproductivity needs to be determined.

References

Barriada JL, Tappin AD, Evans EH, Achterberg EP (2007) Dissolved silver measurements in seawater. Trends Anal Chem 26:809–817
Beck AJ, Sañudo-Wilhelmy SA (2007) Impact of water temperature and dissolved oxygen on copper cycling in an urban estuary. Environ Sci Tech 41:6103–6108
Benn TM, Westerhoff P (2008) Nanoparticle silver released into water from commercially available sock fabrics (vol 42, pg 4133, 2008). Environ Sci Tech 2:7025–7026
Benoit MD, Kudela RM, Flegal AR (2010) Modeled trace element concentrations and partitioning in the San Francisco estuary, based on suspended solids concentration. Environ Sci Tech 44:5956–5963
Bianchini A, Playle RC, Wood CM, Walsh PJ (2005) Mechanism of acute silver toxicity in marine invertebrates. Aquat Toxicol 72:67–82
Blaser SA, Scheringer M, MacLeod M, Hungerbuehler K (2008) Estimation of cumulative aquatic exposure and risk due to silver: contribution of nano-functionalized plastics and textiles. Sci Total Environ 390:396–409
Bloom NS, Crecelius EA (1984) Determination of silver in sea water by coprecipitation with cobalt pyrrolidinedithiocarbamate and Zeeman graphite-furnace atomic absorption spectrometry. Anal Chim Acta 156:139–145
Bloom NS, Crecelius EA (1987) Distribution of silver, mercury, lead, copper and cadmium in central Puget Sound sediments. Mar Chem 21:377–390
Bone AJ, Colman BP, Gondikas AP, Newton KM, Harrold KH, Cory RM, Unrine JM, Klaine SJ, Matson CW, Di Giulio RT (2012) Biotic and abiotic interactions in aquatic microcosms determine fate and toxicity of Ag nanoparticles: Part 2-toxicity and Ag speciation. Environ Sci Tech 46:6925–6933
Böning P, Brumsack H-J, Schnetger B, Grunwald M (2009) Trace element signatures of Chilean upwelling sediments at ~36°S. Mar Geol 259:112–121
Böning P, Brumsack HJ, Bottcher ME, Schnetger B, Kriete C, Kallmeyer J, Borchers SL (2004) Geochemistry of Peruvian near-surface sediments. Geochim Cosmochim Acta 68:4429–4451

Böning P, Cuypers S, Grunwald M, Schnetger B, Brumsack H-J (2005) Geochemical characteristics of Chilean upwelling sediments at ~36°S. Mar Geol 220:1–21

Borchers SL, Schnetger B, Boning P, Brumsack HJ (2005) Geochemical signatures of the Namibian diatom belt: perennial upwelling and intermittent anoxia. Geochem Geophys Geosyst 6:Q06006

Bradford A, Handy RD, Readman JW, Atfield A, Muehling M (2009) Impact of silver nanoparticle contamination on the genetic diversity of natural bacterial assemblages in estuarine sediments. Environ Sci Tech 43:4530–4536

Bruland KW, Franks RP, Knauer GA, Martin JH (1979) Sampling and analytical methods for the determination of copper, cadmium, zinc, and nickel at the nanogram per liter level in sea water. Anal Chim Acta 105:233–245

Bruland KW, Lohan MC (2004) The control of trace metals in seawater. In: Elderfield H (ed) The Oceans and Marine Geochemistry, vol 6, Treatise on Geochemistry (Eds. H.D. Holland and K.K. Turekian). Elsevier, Amsterdam, Chapter 2

Buck NJ, Gobler CJ, Sañudo-Wilhelmy SA (2005) Dissolved trace element concentrations in the East River – Long Island Sound system: relative importance of autochthonous versus allochthonous sources. Environ Sci Tech 39:3528–3537

Burmester T (2002) Origin and evolution of arthropod hemocyanins and related proteins. J Compar Physiol B Biochem Syst Environ Physiol 172:95–107

Chambers BA, Afrooz ARMN, Bae S, Aich N, Katz L, Saleh NB, Kirisits MJ (2014) Effects of chloride and ionic strength on physical morphology, dissolution, and bacterial toxicity of silver nanoparticles. Environ Sci Tech 48:761–769

Clark L, Gobler C, Sañudo-Wilhelmy S (2006) Spatial and temporal dynamics of dissolved trace metals, organic carbon, mineral nutrients, and phytoplankton in a coastal lagoon: Great South Bay, New York. Estuar Coasts 29:841–854

Cozic A, Viollier E, Chiffoleau J-F, Knoery J, Rozuel E (2008) Interactions between volatile reduced sulfur compounds and metals in the Seine estuary (France). Estuar Coasts 31:1063–1071

Crusius J, Thomson J (2003) Mobility of authigenic rhenium, silver, and selenium during postdepositional oxidation in marine sediments. Geochim Cosmochim Acta 67:265–273

Danielsson L-G, Magnusson B, Westerlund S (1978) An improved metal extraction procedure for the determination of trace metals in sea water by atomic absorption spectrometry with electrothermal atomization. Anal Chim Acta 98:47–57

Douville E, Charlou JL, Oelkers EH, Bienvenu P, Colon CFJ, Donval JP, Fouquet Y, Prieur D, Appriou P (2002) The Rainbow vent fluids (36 degrees 14' N, MAR): the influence of ultramafic rocks and phase separation on trace metal content in Mid-Atlantic Ridge hydrothermal fluids. Chem Geol 184:37–48

Eisler R (2010) Compendium of Trace Metals and Marine Biota. Elsevier, Amsterdam

Engel DW, Sunda WG, Fowler BA (1981) Factors affecting trace metal uptake and toxicity to estuarine organisms 1. Environmental parameters. In: Vern-berg J, Calabrese A, Thurberg FP, Vernberg WB (eds) Biological Monitoring of Marine Pollutants. Academic, New York, NY, pp 127–144

Ettajani H, Amiard-Triquet C, Amiard J-C (1992) Etude expérimentale du transport de deux elements traces (Ag, Cu) dans une chaîne trophique marine: Eau - particules (sediment natural, microalgae) – mollusques filtreurs (*Crossostrea gigas* Thunberg). Water Air Soil Pollut 65:215–236

Fabrega J, Luoma SN, Tyler CR, Galloway TS, Lead JR (2011) Silver nanoparticles: behaviour and effects in the aquatic environment. Environ Int 37:517–531

Fisher NS, Breslin VT, Levandowsky M (1995) Accumulation of silver and lead in estuarine microzooplankton. Mar Ecol Prog Ser 116:207–215

Fisher NS, Wang WX (1998) Trophic transfer of silver to marine herbivores: a review of recent studies. Environ Toxicol Chem 17:562–571

Flegal AR, Brown CL, Squire S, Ross JR, Scelfo GM, Hibdon S (2007) Spatial and temporal variations in silver contamination and toxicity in San Francisco Bay. Environ Res 105:34–52

Flegal AR, Sañudo-Wilhelmy SA (1993) Comparable levels of trace metal contamination in two semienclosed embayments: San Diego Bay and South San Francisco Bay. Environ Sci Tech 27:1934–1936

Flegal AR, Sañudo-Wilhelmy SA, Scelfo GM (1995) Silver in the Eastern Atlantic Ocean. Mar Chem 49:315–320

Folsom TR, Grismore R, Young DR (1970) Long-lived gamma-ray emitting nuclide silver-108 m found in Pacific marine organisms and used for dating. Nature 227:941–943

Gallon C, Ranville MA, Conaway CH, Landing WM, Buck CS, Morton PL, Flegal AR (2011) Asian Industrial lead inputs to the North Pacific evidenced by lead concentrations and isotopic compositions in surface waters and aerosols. Environ Sci Tech 45:9874–9882

GEOTRACES (2006) GEOTRACES science plan: an international study of the marine biogeochemical cycles of trace elements and their isotopes. Page 79. SCOR, Baltimore, MD.

Godfrey LV, Field MP, Sherrell RM (2008) Estuarine distributions of Zr, Hf, and Ag in the Hudson River and the implications for their continental and anthropogenic sources to seawater. Geochem Geophys Geosyst 9:Q12007

He D, Bligh MW, Waite TD (2013) Effects of aggregate structure on the dissolution kinetics of citrate-stabilized silver nanoparticles. Environ Sci Tech 47:9148

Hodge VF, Folsom TR (1972) Estimate of world budget of fallout silver nuclides. Nature 237:98–99

Huerta-Diaz MA, Rivera-Duarte I, Sañudo-Wilhelmy SA, Flegal AR (2007) Comparative distributions of size fractionated metals in pore waters sampled by in situ dialysis and whole-core sediment squeezing: implications for diffusive flux calculations. Appl Geochem 22:2509–2525

James RH, Elderfield H, Palmer MR (1995) The chemistry of hydrothermal fluids from the Broken Spur site, 29°N Mid-Atlantic ridge. Geochim Cosmochim Acta 59:651–659

Kalnejais LH, Martin WR, Bothner MH (2010) The release of dissolved nutrients and metals from coastal sediments due to resuspension. Mar Chem 121:224–235

Kalnejais LH, Martin WR, Signall RP, Bothner MH (2007) Role of sediment resuspension in the remobilization of particulate-phase metals from coastal sediments. Environ Sci Tech 41: 2282–2288

Koide M, Hodge VF, Yang JS, Stallard M, Goldberg EG, Calhoun J, Bertine KK (1986) Some comparative marine chemistries of rhenium, gold, silver and molybdenum. Appl Geochem 1:705–714

Kramer D, Cullen JT, Christian JR, Johnson WK, Pedersen TF (2011) Silver in the subarctic northeast Pacific Ocean: explaining the basin scale distribution of silver. Mar Chem 123:133–142

Levard C, Hotze EM, Lowry GV, Brown GE Jr (2012) Environmental transformations of silver nanoparticles: impact on stability and toxicity. Environ Sci Tech 46:6900–6914

Li H, Turner A, Brown MT (2013) Accumulation of aqueous and nanoparticulate silver by the marine gastropod *Littorina littorea*. Water Air Soil Pollut 224:1–9

Liu J, Hurt RH (2010) Ion release kinetics and particle persistence in aqueous nano-silver colloids. Environ Sci Tech 44:2169–2175

Liu J, Sonshine DA, Shervani S, Hurt RH (2010) Controlled release of biologically active silver from nanosilver surfaces. ACS Nano 4:6903–6913

Long A, Wang WX (2005) Assimilation and bioconcentration of Ag and Cd by the marine black bream after waterborne and dietary metal exposure. Environ Toxicol Chem 24:709–716

Luo Y, Dabek-Zlotorzynska E, Celo V, Muir DCG, Yang L (2010) Accurate and precise determination of silver isotope fractionation in environmental samples by multicollector-ICPMS. Anal Chem 82:3922–3928

Luoma SN (2008) Silver Nanotechnologies and the Environment: old problems or new challenges? Woodrow Wilson International Center for Scholars, Project on Emerging Nanotechnologies. Publication PEN 15. September.

Luoma SN, Ho YB, Bryan GW (1995) Fate, bioavailability and toxicity of silver in estuarine environments. Mar Pollut Bull 31:44–54

Martin JH, Knauer GA, Gordon RM (1983) Silver distributions and fluxes in north-east Pacific waters. Nature 305:306–309

McKay JL, Pedersen TF (2008) The accumulation of silver in marine sediments: a link to biogenic Ba and marine productivity. Global Biogeochem Cycles 22:GB4010

Miller LA, Bruland KW (1995) Organic speciation of silver in marine waters. Environ Sci Tech 29:2616–2621

Morford JL, Kalnejais LH, Helman P, Yen G, Reinard M (2008) Geochemical cycling of silver in marine sediments along an offshore transect. Mar Chem 110:77–88

Murozumi M (1981) Isotope-dilution surface-ionization mass-spectrometry of trace constituents in natural environments and in the Pacific. Bunseki Kagaku 30:S19–S26

Murozumi M, Nakamura S, Suga K (1981) Isotope-dilution surface-ionization mass-spectrometry of silver in environmental materials. Nippon Kagaku Kaishi 1981(3):385–391

Navarro E, Piccapietra F, Wagner B, Marconi F, Kaegi R, Odzak N, Sigg L, Behra R (2008) Toxicity of silver nanoparticles to *Chlamydomonas reinhardtii*. Environ Sci Tech 42:8959–8964

Ndung'u K, Ranville MA, Franks RP, Flegal AR (2006) On-line determination of silver in natural waters by inductively-coupled plasma mass spectrometry: influence of organic matter. Mar Chem 98:109–120

Ndung'u K, Thomas MA, Flegal AR (2001) Silver in the western equatorial and South Atlantic Ocean. Deep-Sea Res II Top Stud Oceanogr 48:2933–2945

Ndung'u K (2011) Dissolved silver in the Baltic Sea. Environ Res 111:45–49

Ng TYT, Wang W-X (2007) Interactions of silver, cadmium, and copper accumulation in green mussels (*Perna viridis*). Environ Toxicol Chem 26:1764–1769

Nriagu JO (1996) A history of global metal pollution. Science 272:223–224

Pedroso MS, Pinho GLL, Rodrigues SC, Bianchini A (2007) Mechanism of acute silver toxicity in the euryhaline copepod *Acartia tonsa*. Aquat Toxicol 82:173–180

Purcell TW, Peters JJ (1998) Sources of silver in the environment. Environ Toxicol Chem 17:539–546

Ranville MA, Cutter GA, Buck CS, Landing WM, Cutter LS, Resing JA, Flegal AR (2010) Aeolian contamination of Se and Ag in the North Pacific from Asian fossil fuel combustion. Environ Sci Tech 44:1587–1593

Ranville MA, Flegal AR (2005) Silver in the North Pacific Ocean. Geochemistry, Geophysics, Geosystems 6:Q03M01.

Ratte HT (1999) Bioaccumulation and toxicity of silver compounds: a review. Environ Toxicol Chem 18:89–108

Ravizza GE, Bothner MH (1996) Osmium isotopes and silver as tracers of anthropogenic metals in sediments from Massachusetts and Cape Cod bays. Geochim Cosmochim Acta 60:2753–2763

Reinfelder JR, Chang SI (1999) Speciation and microalgal bioavailability of inorganic silver. Environ Sci Tech 33:1860–1863

Rivera-Duarte I, Flegal AR (1997) Pore-water silver concentration gradients and benthic fluxes from contaminated sediments of San Francisco Bay, California, U.S.A. Mar Chem 56:15–26

Rivera-Duarte I, Flegal AR, Sañudo-Wilhelmy SA, Véron AJ (1999) Silver in the far North Atlantic Ocean. Deep-Sea Res II Top Stud Oceanogr 46:979–990

Rozan TF, Luther GWI (2009) Voltammetric evidence suggesting Ag speciation is dominated by sulfide complexation in river water. In: Taillefert M, Rozan TF (eds) Environmental Electrochemistry: Analyses of Trace Element Biogeochemistry, vol 811, ACS Symposium Series. ACS, Washington, DC, pp 371–387, Chapter 19

Sañudo-Wilhelmy SA, Flegal AR (1992) Anthropogenic silver in the Southern California Bight: a new tracer of sewage in coastal waters. Environ Sci Tech 26:2147–2151

Sañudo-Wilhelmy SA, Olsen KA, Scelfo JM, Foster TD, Flegal AR (2002) Trace metal distributions off the Antarctic Peninsula in the Weddell Sea. Mar Chem 77:157–170

Sañudo-Wilhelmy SA, Rivera-Duarte I, Russell Flegal A (1996) Distribution of colloidal trace metals in the San Francisco Bay estuary. Geochim Cosmochim Acta 60:4933–4944

Sañudo-Wilhelmy SA, Tovar-Sanchez A, Fisher NS, Flegal AR (2004) Examining dissolved toxic metals in U.S. estuaries. Environ Sci Tech 38:34A–38A

Schönbächler M, Carlson RW, Horan ME, Mock TD, Hauri EH (2007) High precision Ag isotope measurements in geologic materials by multiple-collector ICPMS: an evaluation of dry versus wet plasma. Int J Mass Spectrom 261:183–191

Smith GJ, Flegal AR (1993) Silver in San Francisco Bay estuarine waters. Estuaries 16:547–558

Spinelli GA, Fisher AT, Wheat CG, Tryon MD, Brown KM, Flegal AR (2002) Groundwater seepage into northern San Francisco Bay: implications for dissolved metals budgets. Water Resources Research 310.1029/2001WR000827.

Squire S, Scelfo GM, Revenaugh J, Flegal AR (2002) Decadal trends of silver and lead contamination in San Francisco Bay surface waters. Environ Sci Tech 36:2379–2386

Tappin AD, Barriada JL, Braungardt CB, Evans EH, Patey MD, Achterberg EP (2010) Dissolved silver in European estuarine and coastal waters. Water Res 44:4204–4216

Turner A, Brice D, Brown MT (2012) Interactions of silver nanoparticles with the marine macroalga, *Ulva lactuca*. Ecotoxicology 21:148–154

Turner DR, Whitfield M, Dickson AG (1981) The equilibrium speciation of dissolved components in fresh-water and seawater at 25-degrees-c and 1 atm pressure. Geochim Cosmochim Acta 45:855–881

Unrine JM, Colman BP, Bone AJ, Gondikas AP, Matson CW (2012) Biotic and abiotic interactions in aquatic microcosms determine fate and toxicity of Ag nanoparticles. Part 1. Aggregation and dissolution. Environ Sci Tech 46:6915–6924

Wang J, Wang W-x (2014) Salinity influences on the uptake of silver nanoparticles and silver nitrate by marine medaka (*Oryzias melastigma*). Environ Toxicol Chem 33:632–640

Wang WX (2001) Comparison of metal uptake rate and absorption efficiency in marine bivalves. Environ Toxicol Chem 20:1367–1373

Wen L-S, Santschi PH, Gill GA, Paternostro CL, Lehman RD (1997) Colloidal and particulate silver in river and estuarine waters of Texas. Environ Sci Tech 31:723–731

Xu Y, Wang WX (2004) Silver uptake by a marine diatom and its transfer to the coastal copepod *Acartia spinicauda*. Environ Toxicol Chem 23:682–690

Yang L, Sturgeon RE (2002) On-line determination of silver in seawater and marine sediment by inductively coupled plasma mass spectrometry. J Anal At Spectrom 17:88–93

Yoo H, Lee JS, Lee BG, Lee IT, Schlekat CE, Koh CH, Luoma SN (2004) Uptake pathway for Ag bioaccumulation in three benthic invertebrates exposed to contaminated sediments. Mar Ecol Prog Ser 270:141–152

Zhang Y, Amakawa H, Nozaki Y (2001) Oceanic profiles of dissolved silver: precise measurements in the basins of western North Pacific, Sea of Okhotsk, and the Japan Sea. Mar Chem 75:151–163

Zhang Y, Obata H, Gamo T (2008) Silver in Tokyo Bay estuarine waters and Japanese rivers. J Oceanogr 64:259–265

Zhang Y, Obata H, Nozaki Y (2004) Silver in the Pacific Ocean and the Bering Sea. Geochem J 38:623–633

DDT, Chlordane, Toxaphene and PCB Residues in Newport Bay and Watershed: Assessment of Hazard to Wildlife and Human Health

James L. Byard, Susan C. Paulsen, Ronald S. Tjeerdema, and Deborah Chiavelli

Contents

1	Introduction	50
2	Newport Bay and Watershed	51
	2.1 Location	51
	2.2 Climate/Hydrology	51
	2.3 Land Use	52
	2.4 Water Quality	52
3	DDT	53
	3.1 Levels in the Environment	53
	3.2 TMDL Targets	83
4	Chlordane	137
	4.1 Levels in the Environment	137
	4.2 Benthic Triad Analysis of Impairment	141
5	Toxaphene	146
	5.1 Levels in the Environment	146
	5.2 NAS Fish Guidance to Protect Wildlife	149
	5.3 New York State Sediment Guidance to Protect Wildlife	149

J.L. Byard (✉)
11693 Phelps Hill Road, Nevada City, CA 95959, USA
e-mail: doctoxics@aol.com

S.C. Paulsen
Exponent, 320 Goddard, Suite 200, Irvine, CA 92618, USA

R.S. Tjeerdema
Department of Environmental Toxicology, University of California at Davis, Davis, CA 95616, USA

D. Chiavelli
Quantitative Environmental Analysis, LLC, 305 W. Grand Ave., Suite 300, Montvale, NJ 07645, USA

6	PCBs	150
	6.1 Levels in Sport Fish	150
	6.2 California Sport Fish Guidance to Protect Human Health	152
7	Summary	153
References		155

1 Introduction

DDT (dichlorodiphenyltrichloroethane), chlordane, toxaphene and PCBs (polychlorinated biphenyls) are persistent organochlorine chemicals that can still be widely found in soils and aquatic environments decades after use has been discontinued. Under the Clean Water Act, these chemicals are regulated by a total maximum daily load (TMDL) for each watershed to achieve levels that are not toxic to wildlife or humans. In Newport Bay and Watershed (Orange County, California), the development of TMDLs for these legacy organochlorines has been underway for more than a decade.

In 2002, the United States Environmental Protection Agency, Region IX (US EPA Region IX) promulgated TMDLs for DDT, chlordane, toxaphene and PCBs in Newport Bay and Watershed. The finding of impairment, and therefore the necessity for the TMDLs, was based on certain target concentrations for water, sediment and fish. The Santa Ana Regional Water Quality Control Board (SARWQCB or Regional Board) revised and approved these TMDLs in 2007, with the condition that an independent advisory panel (IAP) of experts review the science underlying the TMDL targets. The IAP was formed by Orange County. They met and considered the TMDLs and the underlying science. They pointed out flaws in the science supporting the TMDLs and recommended developing TMDLs based on site specific food chain bioaccumulation of the legacy organochlorines (IAP 2009). The California State Water Resources Control Board (CSWRCB or State Board) and Region IX of the USEPA have recently approved the 2007 TMDLs; implementation of the TMDLs and reconsideration of their targets based on the IAP recommendations was begun in 2013.

The regulated community has pointed out throughout the TMDL proceedings that these organochlorines are no longer in use, that residue levels are declining, and that there are no apparent effects on wildlife or human health. They have also pointed out that many of the targets to be implemented are not based on sound science. The important question for all concerned is whether the chosen targets are scientifically sound and whether current levels meet or exceed scientifically sound targets.

Technical reports that address different aspects of the TMDL process or the targets have been written by US EPA Region IX (2002), SARWQCB (2006) and scientists working for the regulated community (Flow Science et al. 2006; Byard 2011, 2012a, b). Scientists representing the regulated community and regulatory agencies have met on numerous occasions and have exchanged comment letters, including comment letters from outside scientists (Daniel Anderson, Donald MacDonald and ten others), in addition to the report from the IAP. In this review, we provide an analysis of the science underlying the organochlorine TMDLs for Newport Bay and Watershed. Since organochlorines are regulated by the TMDL process in many other locations, the analysis herein represents a case study that may have application to other watersheds.

2 Newport Bay and Watershed

2.1 Location

The Newport Bay Watershed is centrally located in Orange County, California. The 154-square mile Watershed includes portions of the cities of Newport Beach, Irvine, Laguna Hills, Lake Forest, Tustin, Orange, Santa Ana, and Costa Mesa. Runoff from the mountains that surround three sides of the watershed drains across the Tustin Plain and enters Upper Newport Bay via San Diego Creek, one of two major tributaries. Peters Canyon Wash, the other tributary, joins San Diego Creek in the City of Irvine. These combined waterways drain 75,520 acres and are major contributors of freshwater and sediment, and associated pollutants to Newport Bay, which includes both Upper and Lower Newport Bay, as shown in Fig. 1.

2.2 Climate/Hydrology

The area has a Mediterranean climate, with short, mild winters and warm dry summers. Ninety percent of the precipitation occurs during November to April with an average rainfall of approximately 13 in. per year. San Diego Creek has a wide range of water

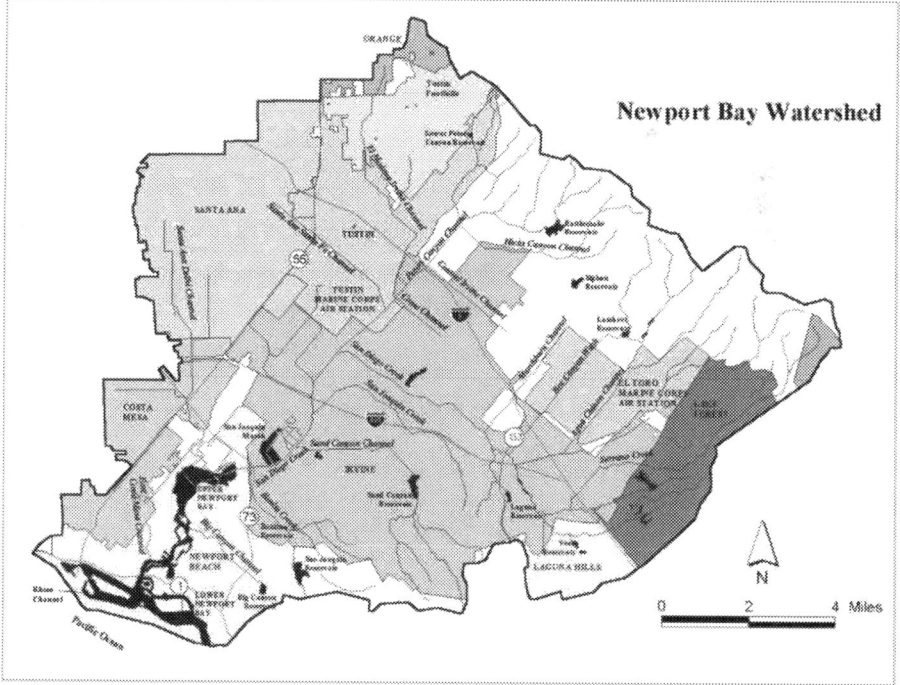

Fig. 1 Newport Bay and Watershed. Figure 1-1 reproduced from U.S. EPA Region IX (2002)

hardness and is influenced by the slightly saline water table (less than 2% salinity). Prior to the 1960s, San Diego Creek was not connected directly to Newport Bay, but an engineered flood control system was constructed within the watershed making the connection permanent. Flows now enter Newport Bay through San Diego Creek throughout the year. San Diego Creek currently has a mean base flow rate of approximately 12 cubic feet second (cfs). Storm events can increase this daily flow to over 9,000 cfs. The Upper Bay's estuary has saline water conditions during dry weather, yet experiences a heavy influx of freshwater from San Diego Creek and other tributaries during major storms. Water quality in the Lower Bay is intermediate between the Upper Bay and the Pacific Ocean.

2.3 Land Use

Land within the Newport Bay Watershed was first settled during the late nineteenth century, leading to the development of agriculture on a large portion of the inland areas. The end of World War II brought urbanization to the area, and land use changed significantly in the past 60 years from agricultural to residential and commercial uses. In 1983, agricultural and urban uses accounted for 22% and 48% of the Newport Bay Watershed, respectively, but by 1993, the proportions were 12% and 64% (US EPA Region IX 1998). As of 2000, agricultural uses had dropped to approximately 7% of the watershed area (US EPA Region IX 2002). Based on plans for development of agricultural lands, this trend is expected to continue.

2.4 Water Quality

Both Upper and Lower Newport Bay and San Diego Creek have been listed as impaired for possessing "unknown toxicity." The toxics TMDL promulgated in 2002 by US EPA Region IX was intended to address the unknown toxicants responsible and develop limits for selenium, metals, organophosphates and organochlorines. Other TMDLs are in place for nutrients (US EPA Region IX 1998), fecal coliforms (SARWQCB 1999), and sediment (SARWQCB 1998).

Researchers have observed acute toxicity in San Diego Creek and Newport Bay. Acute toxicity was observed in urban storm water runoff and in agricultural drainage from some types of crops in the watershed (Lee and Taylor 2001). Bay et al. (2004) collected sediment samples in 2000 and 2001 and found toxicity at multiple locations in both Upper and Lower Newport Bay. It was concluded in both studies that the acute toxicity was not caused by organochlorine compounds, but more likely was attributable to organophosphate (diazinon and chlorpyrifos), carbamate, and pyrethroid pesticides (Lee and Taylor 2001; Bay et al. 2004). Although uses of diazinon and chlorpyrifos have been phased out by the US EPA, other organophosphates, carbamates and pyrethroid pesticides are still used in residential, agricultural and commercial applications, such as the commercial nurseries formerly located in the upper part of the Watershed. Recent studies indicate that the acute toxicity of sediments from Upper Newport Bay is diminishing (Orange County Watersheds 2008–2011).

Understanding sediment loads is important to understanding the fate of organochlorines, because these chemicals bind tightly to soil and sediment particles. Annual sediment loads discharged from the Watershed were estimated at approximately 250,000–275,000 t during the rapid urbanization period of the 1980s and 1990s (US EPA Region IX 1998). Much of the sediment load resulted from in-stream erosion (Trimble 1997). Because of the volume of sediments deposited within the Bay, the Upper Bay was dredged in 1983, 1985, 1988, 1999 and 2010 (Newport Bay Conservancy 2013). Implementation of the sediment TMDL (SARWQCB 1998) has led to reduced sediment loads.

Stream stabilization and other measures have reduced sediment loads to the Bay.[1] Flow rate and suspended sediment discharge samples collected at San Diego Creek showed that although average annual flow volume for the years 2000–2005 was roughly equivalent to the average annual flow volume for 1983–1999, average annual sediment discharge for the latter period was only 42% of the average annual sediment discharge for 1983–1999 (see footnote 1). Orange County's consultant attributed this reduction in sediment load to land development, effectively capping soils, and to erosion control measures in the watershed. Moreover, this consultant found that "[a]s the San Diego Creek watershed becomes further developed, less and less watershed supply of sediment is released during storm events (see footnote 1)."

In the next section, we address the fate of DDT in the Watershed and the science underlying the DDT TMDL targets.

3 DDT

3.1 Levels in the Environment

DDT was first used as an insecticide in California around 1944 and was in wide use by 1947. In 1963, the California Department of Food and Agriculture declared it a restricted material, and 1972 was the last year that DDT was applied to crops in the state (Mischke et al. 1985).

According to the United States Department of Health and Human Services (US DHHS 2002), commercial DDT is a mixture of several congeners, and typically has a composition of 65–80% p,p'-DDT and 15–21% o,p'-DDT. In the environment, DDE and DDD are the major degradation products of DDT. DDT and its congeners are persistent in the environment and have been found in various animal species, in water, in soil, and in sediment (US DHHS 2002). As indicated below and except as otherwise indicated, DDT is used to represent the sum of all measured DDT congeners.

---

[1] County of Orange Resources and Development Department 2006.

DDT is strongly hydrophobic; for example, p,p'-DDE, the main metabolite of concern, has a K_{ow} of 6.956 (de Bruijn et al. 1989). After application to soils, DDT may be lost through both volatilization and biodegradation. Volatilization tends to be the more important removal mechanism initially, while biodegradation is more important later in the removal process (US DHHS 2002). As a result of both these processes, DDT removal from soils tends to be non-linear, and thus the first 50% of DDT tends to be removed from soil more quickly than subsequent halves, such that the half-life of DDT in soil may increase over time (US DHHS 2002).

A variety of studies have been conducted to characterize the half-life of DDT and its metabolites. In temperate climates, the half-life of DDT in soil has been reported to range from 2.3 to 16.7 years (Lichtenstein and Schultz 1959; Racke et al. 1997; Stewart and Chisholm 1971).

Although it has been suggested that other non-organochlorine pesticides, such as dicofol, continue to be used in the watershed and may include small amounts of DDT, the SARWQCB (2006) concluded that dicofol contains minimal levels of DDT and is therefore an "inconsequential continuing source in the watershed." Mischke et al. (1985) concluded that DDT levels in dicofol were too low to account for the DDT soil residues found in their 1985 study of agricultural residues in California soils.

As discussed below, recent data from Newport Bay and Watershed indicate that DDT concentrations are declining in all media where historical and recent data are available, and these data have been used to estimate the half-life of DDT in the local watershed.

3.1.1 Agricultural Soils

Table 1 presents historical DDT concentrations for agricultural soils in the Newport Bay watershed. In general, these soils seem to exhibit a downward trend in DDT concentrations over time, which is expected given a DDT half-life of less than

Table 1 DDT concentrations in agricultural soils in the Newport Bay Watershed

Year	0-12 inch Sample Depth			12-24 inch Sample Depth			>24 inch Sample Depth			Detection Limits (ppm)
	Range of Detected Total DDT (ppm)	Total Samples	Total Non-detect Samples	Range of Detected Total DDT (ppm)	Total Samples	Total Non-detect Samples	Range of Detected Total DDT (ppm)	Total Samples	Total Non-detect Samples	
1985	0.001 - 1.750	12	0							--
1987	0.034 - 1.500	10	0	0.025 - 2.140	10	4	--	10	10	0.016
1988	0.027 - 1.090	10	1	0.095 - 0.150	10	8	0.550 - 0.550	10	9	<0.027 - 0.064
1989	0.024 - 0.791	15	6	0.052 - 0.707	10	5	0.016 - 0.333	19	13	0.016
1990	0.102 - 0.900	4	1	0.110 - 0.910	2	0	0.020 - 0.197	7	3	0.016
1991	0.019 - 0.488	34	4	0.010 - 0.490	32	14				--
1995	0.085 - 0.806	19	1							--
2000	0.007 - 0.132	28	0							0.005
2002	0.005 - 1.620	174	34				0.020 - 0.073	27	17	--
2004	0.002 - 2.000	230	167				0.002 - 0.300	45	36	0.002 - 0.2
2006	0.013 - 0.157	6	1				0.005 - 0.005	6	5	0.005

Sources: Unpublished technical report provided by the SARWQCB (1985); unpublished technical reports provided by The Irvine Company (1985–2006); no data were available for shaded areas and for entries noted with '--'

20 years (Lichtenstein and Schultz 1959; Racke et al. 1997; Stewart and Chisholm 1971) and the fact that DDT use was discontinued in the early 1970s. However, the data reported in Table 1 were not sampled from soils at the same locations. Given that no data were available showing the amounts of DDT historically applied to different areas of the watershed, the DDT data on agricultural soils cannot be used to assess trends over time or local DDT half-life values.

Several agricultural soil DDT data points have also been reported in Mischke et al. (1985) for Orange County. Total DDT concentrations in that report ranged from 0.32 to 2.96 ppm for three different sample locations. However, the precise locations of these samples could not be identified from the report, and thus the data were not useful for establishing trends in agricultural soil DDT concentrations in the Newport Bay watershed.

Peak DDT concentrations at the 12–24 in. depth were generally comparable to concentrations in the top 12 in., while peak DDT concentrations in samples collected from a depth of 24 or more inches were roughly two to sixfold lower than concentrations at the surface. Sampling locations for several sampling years are presented in Fig. 2; note that exact locations for samples collected in 1988, 1991, and 1995 are not known. Data from those years are shown using the average concentration in the approximate location of sample collection.

If one conservatively assumes a half-life of 20 years for DDT in soil, that DDT was banned in 1972, and if other losses or removal mechanisms are excluded, the mass of DDT in the agricultural soils of the Newport Bay watershed would have declined by approximately 71% over the past 36 years solely from soil degradation.

Because DDT adsorbs strongly to soil particles, the predominant pathway for movement in the watershed is via soil erosion. Two related changes within the watershed have served to minimize the transport of DDT to the waters within the Newport Bay watershed. First, urban development initially has led to a conversion of land use away from agricultural use and toward residential, commercial, and industrial development, which increases the impervious land and minimizes direct erosion from land surfaces. Much of this land use conversion has occurred on land that was in agricultural production prior to 1972, when DDT was in use, and much of the land currently in agricultural production was first farmed after 1972, indicating that DDT would not have been applied to these areas. Second, several measures, including channelization and construction of and improvements to the flood control system, have resulted in decreased sediment loads being delivered to the Bay, and therefore, decreased sediment yield of the watershed over time.[2] Development in the watershed has and will continue to reduce the amount of DDT available to biota in the watershed.

---

[2] WRC Consulting Services Inc, Historical Sediment Load Examination, San Diego Creek Watershed. Report prepared for County of Orange, Resources and Development Management Department. June 28, 2006.

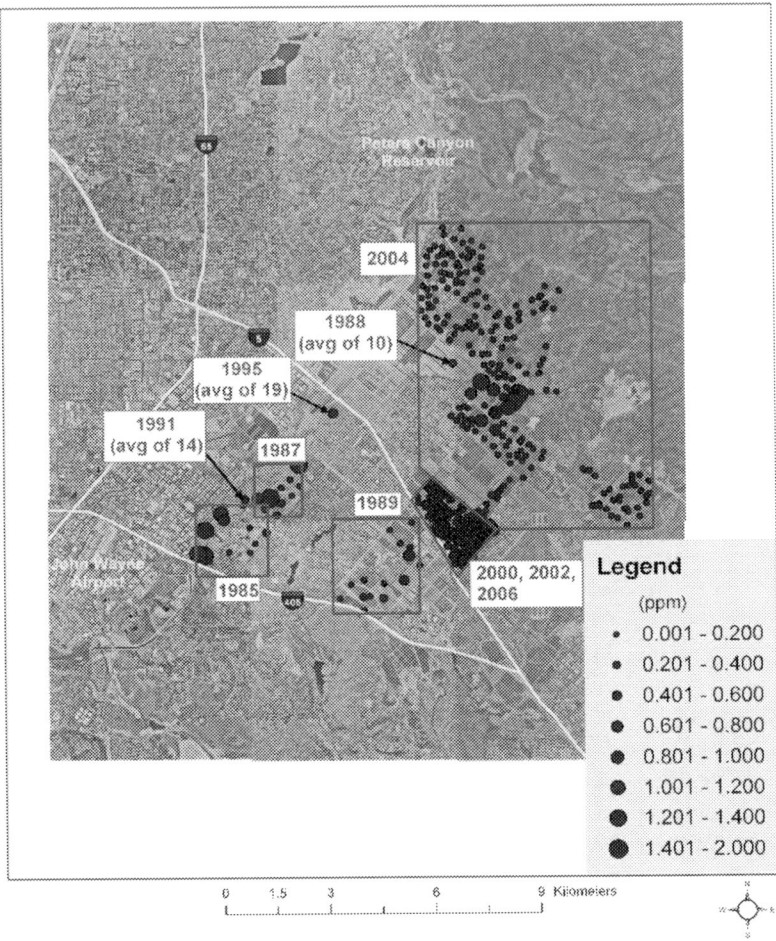

Fig. 2 Locations of agricultural soil samples analyzed for DDT. Data were from unpublished technical reports provided by The Irvine Company (1985–2006). The aerial photo is a composite of photos taken in 1994 and 1995

3.1.2 Sediments

Sediment data are plotted in Fig. 3 for Lower and Upper Newport Bay for the period 1980 through 2011. There are no data available for the period from 1987 to 1995.

Bay-wide trends in sediment DDT concentration over time are difficult to infer from these data for several reasons. First, sampling was conducted by multiple agencies, using multiple methodologies, at varying locations and sample depths. Given this diversity in sampling approach and location, direct comparisons between data from year to year are inappropriate. Second, there is significant movement of sediment into, out of, and within the Bay such that even samples taken in the same

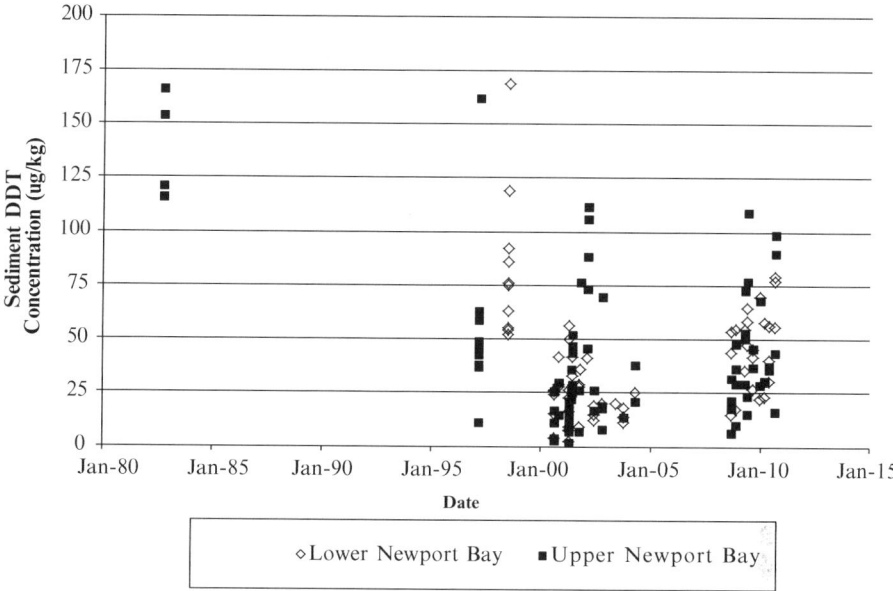

Fig. 3 DDT concentrations in sediment from Newport Bay. Values above detection limits are shown. *Sources*: Orange County PFRD (1980–1986) (Personal communication from Bruce Moore of the Orange County Public Facilities and Resources Department (OCPFRD). Unpublished sediment data for 1980–1986); SCCWRP (1998); Bay et al. (2004); unpublished reports provided by The Irvine Company (2000–2004); US EPA Region IX (2002); Masters and Inman (2000); Orange County Watersheds (2008–2011)

location at two different times may not represent the change in DDT concentration for a specific quantity of sediment. Sediment movement results both from tidal flows and storm flows, as well as from periodic major dredging projects in the Upper Bay, which have occurred in years previously noted. Third, sediment concentrations in Newport Bay may be more indicative of DDT loads from years or decades past, since Bay sediments are transported from the upper watershed in a highly variable, episodic manner, correlated with storm events and wetter-than-average rainfall years.

Thus, DDT concentrations in Bay sediments reflect DDT that was applied many years ago in the upper watershed, and then sorbed to sediments in that location, which were subsequently eroded into a creek channel and transported to the Bay. Finally, Bay sediment DDT concentrations found are not necessarily bioavailable. This is especially true of samples collected from deeper sediment cores. While sample depths were not available for all data plotted in Fig. 3, the Orange County PFRD data from 1980 through 1986 reflect sample depths between 2 and 25 ft, with an average of 11 ft, well below the biologically active layer, which extends only to a depth of approximately 6 in. Thus, these early sediment samples are not indicative of concentrations available to biota in the Bay. For all these reasons, the available sediment data for Newport Bay are not reliable indicators of bioavailable DDT concentration trends in the watershed and should not be used independent of other available data.

Table 2 DDT concentrations in the water column in Newport Bay and Watershed

Date	Water body	Sample location	Fresh water flows	Total DDT (ng/L)[a]
4/23/2001	Lower Bay	Turning Basin	Unspecified	1.29
4/23/2001	Lower Bay	PCH Bridge	Unspecified	1.04
3/12/2002	Rhine Channel	NB3	Unspecified	ND
3/13/2002	Upper Bay	NB10	Unspecified	ND
3/7/2002	San Diego Creek	Campus Drive	Dry weather	ND
3/7/2002	San Diego Creek	Campus Drive	Storm	ND
5/2/2002	San Diego Creek	Campus Drive	Dry weather	ND
5/2/2002	San Diego Creek	Campus Drive	Dry weather	ND
8/12/2002	San Diego Creek	Campus Drive	Dry weather	ND
8/12/2002	San Diego Creek	Campus Drive	Dry weather	ND
11/8/2002	San Diego Creek	Campus Drive	Storm	3
11/8/2002	San Diego Creek	Campus Drive	Storm	ND

[a]Detection limit = 1.0 ng/L
ND is not detected
Bay et al. (2004) and Bay and Greenstein (2003)

3.1.3 Water Column

Twelve water samples have been collected that characterize DDT concentrations in the waters of the Newport Bay and Watershed, as shown in Table 2. Accurate measurement of the very low levels at which DDT is present in water in the Bay is difficult, and only 3 of 12 data points, which were all collected in 2001 and 2002, were above detection limits. For these reasons, no meaningful trend analysis could be performed on concentrations of DDT in water. The California Toxic Rule (CTR) human health regulatory threshold for DDT in water is 0.00059 µg/L, or 0.59 ng/L (US EPA Region IX 2000).

3.1.4 Fish and Mussels

A rigorous statistical analysis of DDT concentration data was conducted for three different media: fish tissue, mussel tissue, and sediment. This analysis demonstrated that DDT concentrations in red shiners and in mussels collected from San Diego Creek, Upper Newport Bay, and Lower Newport Bay are declining in the watershed, and that these trends are statistically significant. DDT concentrations in seven other fish species (for which too few data are available to conduct a robust statistical analysis) are consistent with the trends observed in red shiners and mussels. The likelihood of 11 independent data sets showing a declining trend if a downward trend did not in fact exist is 1 in 2^9—i.e., vanishingly small.

Trends in DDT concentrations are evident in data collected for approximately 20 years in Newport Bay and Watershed. In the case of the fresh water fish species red shiner, tissue DDT concentration data are available from 1983 through 2002 (n = 54);

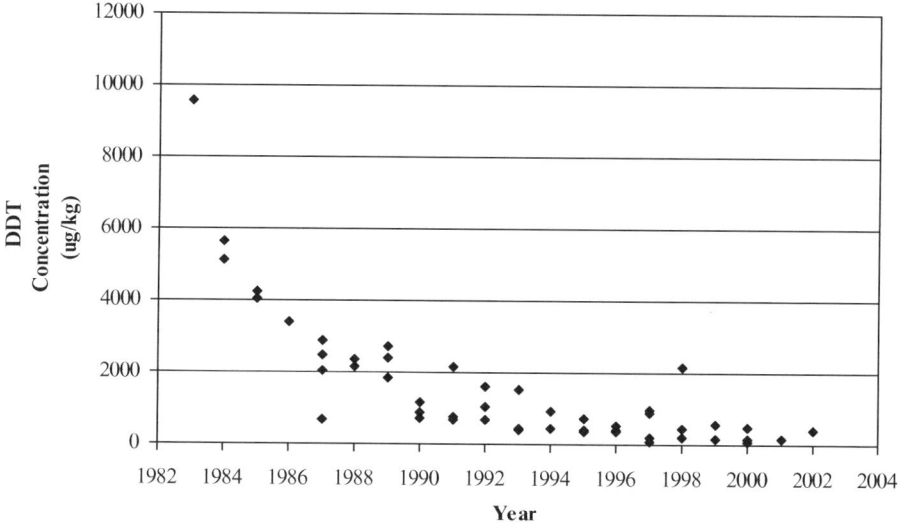

Fig. 4 DDT concentrations in red shiners from San Diego Creek and tributaries. Data from California Toxic Substances Monitoring Program (TSMP 1983–2002). Red shiner data are not available after 2002

the data are plotted in Fig. 4. Concentrations of DDT in red shiner are strongly indicative of concentrations within the Watershed, as this species has a short life span of approximately of 2 years (Baird and Girard 1853) and residents do not range outside of the fresh water streams flowing into Newport Bay.

Exponential regression was used to evaluate the strength of the declining trend in DDT concentration in red shiner tissue over time. A regression through the entire dataset indicated a highly significant exponential decline in DDT concentrations in red shiner tissue. The calculated rate of decline (without outliers) is −0.183 per year (equivalent to a DDT half-life in the watershed, as calculated from the surrogate endpoint of red shiner tissue, of 3.8 years), significantly shorter than the 20-year half-life typical for DDT decay in terrestrial soil. To confirm these trends, a regression analysis was performed for two 10-year sub-periods within the data set, 1983–1992 and 1993–2002, to evaluate whether rates of DDT loss in this species have changed over time. The rate of decline of DDT concentration in red shiners was lower for the later period (−0.135 per year) than for the earlier period (−0.245 per year), but both rates are highly significant.

Trends in DDT concentrations were also evaluated for seven additional fish species (California killifish, spotted sand bass, California halibut, diamond turbot, black perch, striped mullet, and yellow fin croaker) for which three or more DDT concentration data points were available during a time range of five or more years up until 2004. Although the data sets for any one of the additional fish species, taken alone, contained too few data points to infer long-term trends in Newport Bay, and in spite of possible sources of DDT for these fish outside of the Bay, the combined datasets from these seven species are consistent with the strong trend evident in the red shiner data set. The combined evidence from all fish species lends far more weight to the conclusion that fish tissue DDT concentrations are declining in the watershed than could be concluded from data from any single species considered alone. As to the important point that some of these fish species may range outside the Bay to feed, and thus concentrations of DDT in the tissues of these fish species may reflect DDT obtained outside of Newport Bay, Allen et al. (2004) concluded that "monitoring studies are needed to determine if elevated DDT levels in the popular sport fishes noted above are due to contamination in the bay or to sources outside the bay." Also, in a follow-up study, Allen et al. (2008) found mean DDT levels in striped mullet from Newport Bay to be 499 ppb. As in the 2004 study (Allen et al. 2004), the authors caution that: "striped mullet (the species with the highest DDT levels) can leave the Bay and move up and down the coast. Its very high levels of DDT may be accumulated outside the Bay (perhaps in the Los Angeles/Long Beach Harbor area,"

Allen et al. (2008) also reported whole-body DDT levels in seven species of forage fish that are consumed by birds in Newport Bay. The species average DDT level was 143 ppb. The mean level was skewed by one species, the deep body anchovy, a nocturnal feeder that had an average whole-body DDT level of 495 ppb.

Like the red shiner, mussels are good bioindicators. Mussel tissue data from three locations in the Newport Bay Watershed—San Diego Creek, Upper Newport Bay, and Lower Newport Bay—were evaluated for trends in DDT concentrations over time. Like red shiner data, mussel tissue data collected since 1982 show statistically significant declines in DDT concentrations (Fig. 5). An exponential regression analysis of mussel data, including the entire period of record from 1982 to 1999,

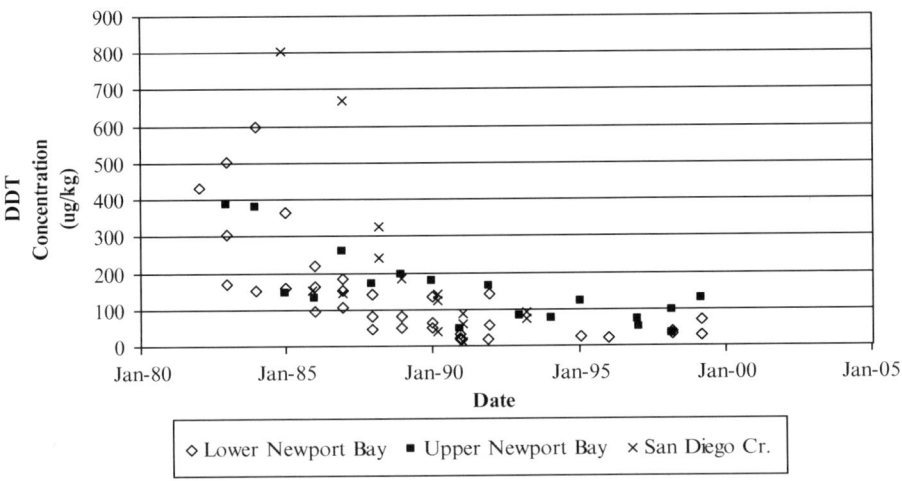

Fig. 5 DDT in mussels from Newport Bay and Watershed. Data from California Mussel Watch Program (1980–2000). The mussel watch program ended in 2000

showed a significant DDT concentration decline rate in mussels both when all three locations were considered together (−0.133 per year, equivalent to a half-life of 5.2 years), and for each individual location (San Diego Creek=−0.292 per year; Upper Newport Bay=−0.095 per year; Lower Newport Bay=−0.156 per year). Mussel data were also analyzed for the periods 1982–1990 and 1991–1999. The rate of decline of DDT concentrations in mussel tissue was statistically significant only for the earlier period (−0.236 per year). The rate of decline was lower for the later period, but period regressions for the later period have low statistical power due, in part, to small sample size. When the entire mussel tissue data set is considered, statistically significant rates of decline in DDT concentration are evident for each of the three locations, and these trends are consistent with trends observed in the fish tissue data sets.

However, as noted below, measured concentrations of DDT in red shiner and mussels indicate a half-life for DDT in biota of the watershed of 3.8 years and 5.2 years, respectively. These observed half-lives suggest that the fraction of DDT removed from the watershed as a whole since 1972 may be significantly higher than 70%, and declining concentrations of DDT in biota are likely from a combination of factors in addition to loss from soils.

3.1.5 DDT in Birds

Birds are exposed to DDT through their diet. Fish-eating birds typically attain the highest levels. DDT levels in fresh eggs correlate with the most sensitive toxic endpoint, eggshell thinning to the point of shell breakage, hatching failure and lowered productivity.

Limited data are available on DDT levels in bird tissue from Newport Bay and Watershed. In a 1984 internal memo, Harry Ohlendorf of the National Wildlife Service reported organochlorine levels in salvaged eggs from the endangered light-footed clapper rail; the eggs were collected during the period 1979–1981.[3] DDE levels ranged from 0.34 to 9.6 ppm. More recently, Sutula et al. (2005) reported total DDT levels of 0.45–1.07 ppm in nonviable eggs collected from light-footed clapper rail nests. The egg with the highest level had the thinnest shell. However, these more recent levels were less than those reported by Goodbred et al. (1996) for light-footed clapper rail eggs from the nearby Tijuana Slough, where eggshell thinning was not observed.

In 2004 and 2005, Gary Santolo at CH2M Hill collected eggs from six species of birds from Newport Bay and Watershed. DDE levels and shell thickness were measured in the eggs. Although the study has not as yet been published in the open literature, Santolo communicated the following general conclusions[4]: no correlation was found between shell thickness and DDE levels in American coot, black skimmer, Forster's tern, killdeer, or black-necked stilt eggs. However, there was a negative

---

[3] Richard Zembal, personal communication.
[4] Gary Santolo, personal communication.

correlation between DDE levels and shell thickness in an analysis of American avocet eggs, although the shell thinning was not within the range that would result in hatching failure.

The recent successful breeding of two osprey pair in Newport Bay implies that levels of DDT in the fish diet of this sensitive species are nontoxic. The breeding data are presented and discussed later in this review.

Many bird species resident to Newport Bay and Watershed have been studied in regard to the levels and effects of DDT at other locations. A detailed review of the effects of DDT on shell thinning and hatching success in the brown pelican, osprey, petrels, and sparrow hawk can be found in later chapters of this review, where effects from DDT in these species played a central role in establishing guidance levels for fish and water. The endangered species, the Belding's savannah sparrow and California gnatcatcher, are not known to be sensitive to the reproductive effects of DDT, most likely because these species feed in a food chain with a relatively low potential for bioaccumulation. The American coot (*Fulica americana*), eating mostly a vegetable diet, would also not be expected to be sensitive to the reproductive effects of DDT. The peregrine falcon is highly sensitive to the reproductive effects of DDT when consuming birds feeding in an aquatic environment. However, the lack of nesting sites (remote rock cliffs) proximate to Newport Bay and Watershed would preclude this food pathway. A possible exception is nesting on bridges and buildings. The following is a review of representative scientific studies of the levels in and effects of DDT on some additional bird species found in Newport Bay and Watershed.

DDT and Terns. There are many reports on DDT levels in tern eggs as well as associated measures of eggshell thickness, hatching success and productivity. Data from Forster's tern, common tern, Caspian tern and least tern are summarized below. The terns are closely related, providing a measure of susceptibility to DDT effects on reproduction for the genus and for the threatened tern species, the least tern.

King et al. (1991) measured DDE levels, eggshell thickness, and hatching success in Forster's tern (*Sterna forsteri*) eggs from two Texas bays. Seventy-one eggs were analyzed for DDE. One egg was taken per nest. Hatching success was monitored in the remaining eggs. DDE levels in eggs ranged from 0.1 to 9.0 ppm. The geometric mean concentrations of DDE in eggs from the two bays were 0.8 and 1.6 ppm. No correlation was found between the level of DDE in eggs and eggshell thickness or hatching success.

Ohlendorf et al. (1988) measured DDT levels in ten Forster's tern eggs from Bair Island in San Francisco Bay. The geometric mean concentration of DDT (all DDE) was 1.92 ppm with a range of 0.88–7.1 ppm. DDE levels were not correlated with eggshell thickness.

Vermeer and Reynolds (1970) reported DDE levels ranging from 2.0 to 25.2 ppm in ten egg composites from common terns (*Sterna hirundo*) in Central Canada in 1968–1969 at the height of the DDT era.

Switzer et al. (1971) reported a mean DDE level of 7.57 ppm in 68 eggs collected from a common tern colony in Alberta, Canada in 1969. DDE levels ranged from 0.64 to 104 ppm. Productivity of the colony was very low, but the authors found no correlation between DDE egg levels and eggshell thickness. A follow-up study in

1970 (Switzer et al. 1973) found higher productivity and lower levels of DDE. Mean DDE levels were 4.52 ppm with a range of 0.13–26.2 ppm. A weak negative correlation was found between eggshell thickness and DDE level.

Fox (1976) reported DDE levels in eggs from common terns nesting in Alberta, Canada. Thirty-nine intact eggs had a mean concentration of 3.42 ppm DDE. A pooled sample of five eggs with dented shells had a mean concentration of 6.67 ppm DDE. The average shell thickness of intact eggs was not different from common tern eggs collected prior to the DDT era; average shell thickness in the dented eggs was 12.5% thinner. One could conclude, based on these findings, that the threshold for egg shell thinning and hatching failure for common terns is greater than 3.42 ppm.

Nisbet and Reynolds (1984) also reported high levels of DDE in a small sample of common tern eggs that failed to hatch during the 1975 breeding season. DDE levels in the three highest eggs ranged from 1.8 to 4.6 ppm. However, in 1971–1972, when the highest levels of DDE were measured in their study, productivity was very high.

Ohlendorf et al. (1985) concluded from the work of Fox (1976) and Switzer et al. (1973) that: "In Common Terns, DDE contamination above 4 ppm was thought responsible for reduced eggshell thickness and quality and lowered hatching success." Since DDE levels range widely above and below the mean of 4 ppm and hatching failure only occurs in a fraction of the eggs, the threshold for hatching failure may be well above 4 ppm.

Weseloh et al. (1989) concluded that DDE was no longer an important factor in the population dynamics of common terns in the Great Lakes. Geometric mean DDE levels ranged from 0.95 to 2.46 ppm at four locations. Eggshell thickness did not correlate with DDE level.

Hoffman et al. (1993) measured geometric mean levels of 1.7–2.9 ppm DDE in eggs from four colonies of common terns in the Great Lakes. The DDE levels ranged from 0.60 to 5.0 ppm. Embryotoxicity observed in the eggs was attributed to PCBs and dioxins. DDE in the eggs was not considered to be sufficiently high to singly account for the observed embryotoxicity.

King et al. (1978) reported mean DDT levels in Caspian tern (*Sterna caspia*) eggs of 15.1 ppm, but no correlation between DDT level and eggshell thickness.

Struger and Weseloh (1985) reported mean DDE levels of 4.6–8.8 ppm in Caspian tern eggs collected in 1980 at various locations in Lake Michigan. Eggshells were at or above the thickness measured in eggs collected prior to the DDT era. The authors concluded that: "Organochlorine levels exhibited in Lake Michigan in 1980 do not appear to have had a detrimental effect on reproductive success in Caspian Terns."

Ohlendorf et al. (1985) reported a geometric mean level of 9.3 ppm DDE (ranging from 2.1 to 56 ppm) in Caspian tern eggs collected in San Diego Bay in 1981. More than one-third of the eggs were lost before hatching, mostly from failure to hatch. On average, eggshells were 7.8% thinner. The authors stated that: "We suspect that higher DDE concentrations in eggs from some nests were, at least in part, responsible for reduced hatching success." Since shell thinning and hatching failure are correlated with DDE levels in eggs, one would expect that the more than one-third hatching failure occurred in eggs with DDE levels above the geometric mean of 9.3 ppm.

Ohlendorf et al. (1988) reported geometric mean DDE levels of 6.93 ppm in 22 Caspian tern eggs from San Francisco Bay and 7.64 ppm in ten Caspian tern eggs from Elkhorn Slough in 1982. In comparing these levels with levels from other reports, the authors stated that: "Caspian Terns in the Great Lakes (Struger and Weseloh 1985) experienced good reproductive success when mean DDE concentrations in eggs were generally similar to—and PCBs generally higher than—concentrations we found in California."

Blus and Prouty (1979) concluded that geometric mean DDE levels in 44 least tern eggs (*Sterna antillarum browni*) from South Carolina in 1972–1975 were low (0.33–0.63 ppm with a range of 0.19–1.22 ppm) and posed no identifiable threat to the species. Eggshells were 2–7% thinner than shells of eggs collected prior to the DDT era.

Hothem and Zador (1995) reported an overall geometric mean of 0.936 ppm DDE in nonviable least tern eggs collected from San Diego Bay in the mid-1980s. Factors such as putrefaction and desiccation in nonviable eggs do not cause loss of DDT, but can produce changes in fresh weight, and therefore the concentration of DDT. Hothem and Powell (2000) sampled 14 least tern eggs from three locations in the San Diego area. The eggs were taken after the breeding season. Although not stated, these eggs also were most likely nonviable. Geometric mean concentrations of DDE in eggs from the three sites ranged from 0.23 to 0.56 ppm. Hothem and Powell (2000) concluded as follows: "Similar or higher mean concentrations in least terns (0.19–1.22 µg/g) from South Carolina (Blus and Prouty 1979) and in Forster's terns from Texas (1.6 µg/g) (King et al. 1991) were not thought to pose a threat to reproduction. Likewise, DDE should not pose a threat to either species in our study."

DDT and Cormorants. Eleven cormorant colonies were studied by Anderson et al. (1969) in the upper midwest and central Canadian provinces in 1965. DDE residue levels were as high as 45 ppm in cormorant eggs with an average of 10.4 ppm. Egg size, weight and thickness varied between the locations. Egg laying is a mechanism for excretion of DDT. Egg residues are more closely related to residues stored in lipid than recent dietary intake. Eggshell thickness was decreased 8.3%. Increases in shell thickness during rebreeding suggest that low DDT levels in local diets were more important than reductions in DDT from utilization of lipid stores during breeding. One population of cormorants, with a 25% decline in eggshell thickness, had recently decreased to nearly zero. The authors claim that the eggshell thinning-DDE regression is linear to zero concentration of DDE. A minimal effect level could not be established. Figure 6 illustrates the eggshell thinning dose-response in cormorants.

Faber and Hickey (1973) reported on a 1969–1970 survey of egg residues and eggshell thinning in fish-eating birds from the upper Great Lakes states and Louisiana. The authors suggest that significant decreases in shell thickness will be found in virtually all fish-eating birds in these parts of America. "We are uncertain about the biological significance of decreases in shell thickness below 10%. Certainly, widespread eggshell breakage does not occur with changes below this magnitude." The level of DDE residue necessary to cause eggshell thinning varies greatly among species. Double-crested cormorants developed eggshell thinning of approximately 12% at 17.5 ppm total organochlorine residues.

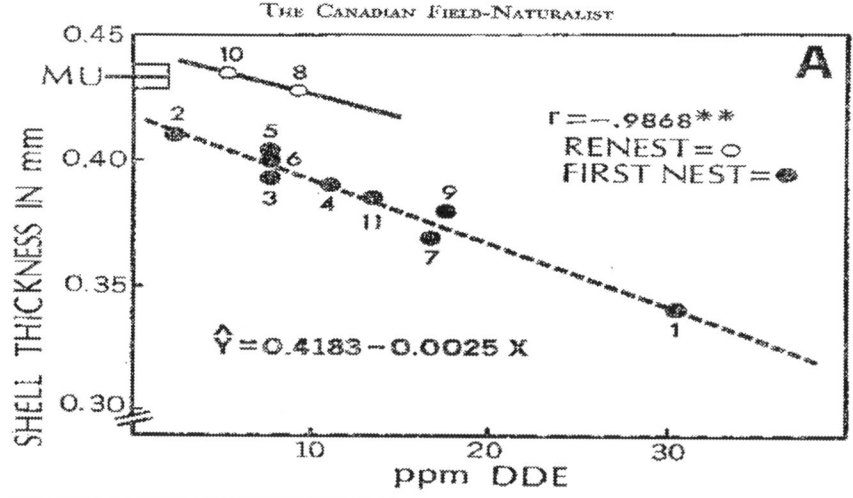

Fig. 6 Eggshell thinning dose response in cormorants. Figure 3 in Anderson et al. (1969) reproduced with permission

Gress et al. (1973) reported on a survey of double-crested cormorant breeding colonies in the Channel Islands and the islands off of the west coast of Baja, California in 1969–1972. Breeding was almost nonexistent in colonies on the Channel Islands and South Los Coronados Island. Breeding appeared unaffected on San Martin Island farther south. No crushed eggs were found on San Martin. Eggshell thinning was 29% and 38% on Anacapa and Los Coronados, respectively. Gress stated that "The San Martin eggshells show no significant differences of any of the parameters from the museum specimens." DDE residues in eggs were 32 ppm, 24 ppm and 1.7 ppm on Anacapa, Los Coronados and San Martin, respectively. He noted that other studies on double-crested cormorants had not found reproductive impairment with DDE residues as high as 10.4 ppm. DDE was associated with 8.3% eggshell thinning. He concluded that "The comparatively low levels of DDE reported suggest that the degree of thinning, if present, would not be sufficiently great to affect reproductive success." Comparisons with studies of interior populations indicated that the relationship between DDE residues and eggshell thinning were the same. In addition, 80% of the variation in eggshell thickness could be explained by the regression on the natural log of DDE. The 1972 survey suggested that both the brown pelican and double-crested cormorant were beginning to recover.

The recovery was attributed to the fact that the DDT manufacturing plant in Los Angeles stopped discharging wastes to the Los Angeles outfall in April, 1970.

Morrison et al. (1978) reported on DDE residues and shell thickness in cormorant eggs collected in Texas in 1976–1977. The results were compared with an earlier study by King, in which cormorant eggs were collected in 1970. The results of the King study were provided by personal communication to the authors from K. A. King. DDE residues had declined dramatically in cormorant eggs from 1970 to 1976–1977 (Table 3).

Eggshell thickness was not significantly affected in either the 1970 or 1976–1977 studies, although the latter shells were thicker (Table 4).

The authors concluded that there was little difference in thickness between the pre-DDT era shells, the 1970 shells and the 1976–1977 shells. "Most authors agree that a 10–20% change in shell thickness is needed before reproductive failures are indicated." and "Cormorant eggshell thickness was apparently not affected by residues in the 1970s in Texas."

Pearce et al. (1979) reported DDE residues in cormorant eggs collected along eastern Canadian coastal waters from 1970 to 1976. Average residues by site ranged from 1.49 to 8.57 ppm. Individual eggs contained from 0.16 to 20 ppm DDE. The authors reported measuring shell thickness, but no data were present in the publication. The authors claimed that 10 ppm DDE in eggs produces 20% shell thinning. This conclusion was based on an extrapolation of the residue—shell thinning data. Again, no data or regression plots were present in the article.

Table 3 Marked reduction in DDE in cormorant eggs from Texas from 1970 to 1976–1977. Data from Table 1 in Morrison et al. (1978)

Residues in olivaceous cormorant eggs in Texas[a]							
	1970 (n=5)			1976–1977 (n=7)			
Residue	Mean	S. E.	(%)	Mean	S. E.	(%)	% Change
p,p'-DDE	6.22	2.08	100	0.400	0.036	100[b]	−93.6

[a] Values represent residues on a wet-wt basis
[b] $p<0.05$

Table 4 Shell thickness in cormorant eggs collected in Texas. Table 2 in Morrison et al. (1978) reproduced with permission

TABLE 2
SHELL THICKNESS OF OLIVACEOUS CORMORANT EGGS IN TEXAS (MM)

				% Change from	
Date	n (eggs)	$\bar{x}$	S.E.	Pre-1940	1970
Pre-1940	75	0.328	0.004	—	—
1970	24	0.323	0.006	−1.5	—
1976–77	21	0.341	0.004	+4.0*	+5.5*

* $p < 0.05$, t-test.

Weseloh et al. (1983) reported on the status of double-crested cormorant colonies in Lake Huron. Six colonies were studied in 1972 and 1973. DDE residues in eggs averaged 14.5 ppm. Eggshell thickness was reduced an average of 23.9%. Egg breakage, hatching failure, and population declines were evident.

Fossi et al. (1984) reported high levels of DDE in cormorant eggs collected from the Danube Delta. DDE levels in 13 eggs averaged 9 ppm. Eggshell thickness was not measured. The authors noted that: "Despite the heavy contamination of the eggs, however, the population of the colonies of Common Cormorant seems to have stabilized...".

King and Krynitsky (1986) studied cormorants nesting in Galveston Bay from 1980 to 1982. DDE levels in eggs averaged 1.73 ppm in 1980 and 0.67 ppm in 1981. Mean shell thickness for the period 1980–1982 was similar to eggs collected prior to the DDT era. Eggs collected from Galveston Bay in 1970 (King et al. 1978) were 7% thinner; eggs collected in 1980 were 5% thinner; eggs collected in 1981 were 3% thinner; eggs collected in 1982 were 1% thicker. The 3% and 1% effects were not statistically significant. One eggshell from an egg collected in 1980 was 22% thinner than in pre DDT era eggs. Although not indicated by the authors, this egg may have contained the highest residue measured in the 1980 eggs ($N=13$). That level was 31 ppm DDE. The authors noted that cormorant populations had remained stable in recent years.

Dirksen et al. (1995) reported a detailed study of organochlorines in cormorants in the Netherlands. Reproductive effects of DDE were confounded by high levels of PCBs in adult tissue and eggs. However, the authors concluded that 4 ppm DDE in cormorant eggs produced 5% shell thinning. They also noted that the threshold for population reproductive failure and population instability was associated with shell thinning of 20%. This level of thinning was associated with egg residues of 10 ppm.

In 1998, the US Department of the Interior published a National Irrigation Water Quality Program Information Report No. 3 titled: "Guidelines for Interpretation of the Biological Effects of Selected Constituents in Biota, Water, and Sediment. DDT." The US Fish and Wildlife Service participated and presumably wrote the section on toxicity to avian species. According to the report, "Toxic effect levels for various types of birds are presented in Table 16." Beginning on page 70, Table 16 lists various avian species, the DDTs studied, the concentration in eggs, the effects observed, and the reference. For the double-crested cormorant, a concentration of 10 ppm of DDE in eggs was stated to cause 20% shell thinning. The reference for this data point is the Pearce et al. (1979) article discussed above. This study claims to have measured shell thinning and to have correlated the shell thinning with DDE residues. However, no shell thinning data and regression plots are to be found in the publication. Hence, this data point in the Department of Interior review is based only on a statement without data or analysis. Comparison of other data points in Table 16 with the referenced article revealed errors and misinterpretations.

For example, Table 16 lists 1 ppm DDE in Western grebe eggs as causing 1% shell thinning. The DDE concentration reported in the cited study was 1.4 ppm, not 1 ppm (Boellstorff et al. 1985). The 1% was reported by Boellstorff et al. (1985) to not be statistically significant. The authors concluded: "Thickness of grebe eggshells collected at Tule Lake NWR in 1972 and 1981 and in northern California from 1952 to 1960 were not significantly different from each other and were not thinner than eggs collected before 1947 (Table 4)."

The very next line in Table 16 states that 5.4 ppm DDE caused 2.3% eggshell thinning and reduced productivity. The research article cited for this data point (Lindvall and Low 1980) reported a DDE residue of 6.6 ppm and a thinning of 3.1%. The authors did not conclude that productivity was reduced. To the contrary, the authors concluded: "The small amount of eggshell thinning seen in western grebe eggshells at Bear River MBR appeared to have little or no effect on reproduction, because no crushed, cracked, or broken eggs were seen during this study. Average brood sizes of 1.6 in 1973 and 1.8 in 1974 from Bear River compare well with the Rudd and Herman determination of a normally reproducing population (18)."

The Department of the Interior report also states in Table 16 that less than 1 ppm DDE produced 6.5% shell thinning in black-crowned night-herons. The reference for this data point (Findholt and Trost 1985) reported a linear regression of shell thickness and log DDE egg residue that had a zero residue intercept of 0.26 mm. Since pre-DDT era shells in this study were 0.275 mm, the linear regression is likely to be inaccurate, particularly at low residue levels. A similar phenomenon has been reported in brown pelican studies. The obvious fallacy in the Table 16 listing is made clear by the fact that eggs containing 1.01–4.0 ppm DDE had thicker shells than eggs with less than 1 ppm DDE.

Table 16 states that 0.52 ppm DDE in common goldeneye eggs causes 15.4% shell thinning and egg breakage. The 15.4% shell thinning is a comparison of 1981 Minnesota colonies with North Dakota and Manitoba eggs collected in 1896 and 1903. The authors (Zicus et al. 1988) conclusion on egg breakage is as follows: "The high rate of egg breakage observed for Common Goldeneyes may be related to eggshell thinning or may be characteristic of the species and perhaps a result of frequent nest parasitism."

Finally, Table 16 states that 12 ppm DDE in Leach's storm petrel eggs results in 12% eggshell thinning. The cited reference (Noble and Elliott 1990) reports only on raptors and makes no mention of Leach's storm petrel. In the few data points in Table 16 that were checked against the original publications, the Department of Interior report repeatedly made errors and misrepresentations of the literature findings on the effects of DDT on avian reproduction.

Custer et al. (1999) reported on cormorant colonies on Cat Island in Green Bay, Wisconsin. Eggs contained 3.9 ppm DDE and 13.6 ppm PCBs. DDE concentration correlated with decreased shell thickness and hatching failure (thinning data were not reported). However, the authors concluded that reproductive performance was generally good to excellent compared to other locations, including those considered to have low levels of persistent organochlorine contamination. "Number of young produced (2.0–2.3 to 12 days of age) was also similar or greater than the 0.7–2.5 young per nest reported in relatively uncontaminated colonies." "…DDE-contamination does not seem to be a significant risk factor to double-crested cormorant populations in this region." A low level of chick deformities was not attributed to DDE.

Cormorants are less sensitive to the reproductive effects of DDTs than ospreys and brown pelicans. Residues of DDE in eggs in excess of 10 ppm, resulting in eggshell thinning of 15% or more, appear to be necessary to produce significant hatching failure.

DDT and Black Skimmers. The black skimmer (*Rynchops niger*) feeds in estuaries, catching small fish by skimming the water surface with its lower mandible; maximum lifespan is 20 years (US GS 2010). King et al. (1978) reported 4% eggshell thinning in black skimmer eggs collected in 1970. Mean DDT levels in the eggs were 9.68 ppm. No correlation was found between DDT levels and eggshell thickness.

Blus and Stafford (1980) reported that DDE ranged from 0.81 to 12.1 ppm in eggs from black skimmer nests that apparently failed. Eggs from successful nests contained 0.43–3.40 ppm. Overall eggshell thinning was 5%. The authors concluded that DDE and other pollutants had little influence on overall productivity of black skimmers in South Carolina.

White et al. (1984) reported that 35% of black skimmer eggs collected in 1978–1981 along the Texas coast contained DDE levels of 10 ppm or more (levels of DDE ranged as high as 51 ppm), but no significant correlation was found between residue levels and fledgling success. The authors state that: "Some degree of eggshell thinning was detected in most colonies (9/10), ranging from 4% to 12%, but thinning was below that (15–20%) believed to cause population declines in other avian species (Anderson and Hickey 1972; Blus et al. 1972a; Longcore and Stendell 1977)."

King and Krynitsky (1986) assessed DDT residues, shell thinning and reproductive success in black skimmers along the Texas coast in 1980–1982. Geometric mean DDE levels in eggs were 1.62–3.25 ppm (range of 0.2–86 ppm). Eggshell thinning for the years 1980, 1981 and 1982 occurred to the extent of 2–6%. Some cracked or crushed and broken eggs were observed to be up to 36% thinner than shells of eggs collected prior to the DDT era. Overall, DDE level did not correlate with shell thickness.

Custer and Mitchell (1987) reported on a study of black skimmers along the Texas coast in 1984. Geometric mean DDE levels were 5.9 ppm (range of 2.3–17.9 ppm) in eggs from nests where the remaining eggs did not hatch and 1.9 ppm (range of 0–7.4 ppm) in eggs from nests where all of the remaining eggs hatched. DDE level did not correlate with shell thickness. The authors concluded that: "The breeding population of Black Skimmers in Texas does not seem to be declining nor does DDT contamination seem to be a major influence on skimmer numbers." In a follow-up publication, King et al. (1991) reported that: "we found no evidence that shell thinning of either tern (7%) or skimmer (5%) eggs adversely affected reproduction in 1984."

DDT and Black-necked Stilts. Setmire et al. (1993) collected 84 black-necked stilt (*Himantopus mexicanus*) eggs from the Salton Sea area in 1988–1990. The geometric mean concentration of DDE was 2.57 ppm with a range of 0.05–12.1 ppm. Eggshell thinning was estimated to be 7% at 12 ppm DDE from a plot of DDE egg residue against eggshell thickness from an Imperial Valley data set of 33 black-necked stilt eggs. However the relationship between DDE level and eggshell thinning was not statistically significant.

Henny et al. (1985) reported that 40 black-necked stilt eggs from Carson Lake, Nevada contained from 0.31 to 15.6 ppm DDE. DDE levels did not correlate with eggshell thickness and shell thickness was similar to that of pre-DDT eggs. The authors concluded that: "No significant relationship was detected between DDE and eggshell thickness for stilts." They noted that Morrison and Kiff (1979) had reported only 1.9% eggshell thinning in stilts from Utah in 1959.

Henny et al. (2008) reported on DDE levels in black-necked stilts collected from the Salton Sea area from 1986 to 2004. Geometric mean levels ranged from 0.55 to 2.91 ppm by colony location, with individual values ranging from 0.05 to 23 ppm. The lowest levels were encountered in 2004, with a geometric mean concentration of DDE of 0.55 ppm and a range of 0.19–1.7 ppm. Eggshell thickness was no different than in stilt eggs collected in Utah prior to the DDT era. The authors concluded that: "The DDE concentrations in eggs documented in this and other studies seem not to produce eggshell thinning in stilts and the associated adverse effect on reproductive success."

DDT and the American Avocet. The American avocet (*Recurvirostra americana*) and black-necked stilts are closely related species. Vermeer and Reynolds (1970) reported 3.32 and 3.16 ppm DDE in ten egg composites of avocet eggs collected in central Canada in 1968–1969. Hunt (1969) reported that avocet and killdeer eggs from the Sacramento Valley averaged 13 ppm DDE.

Henny et al. (2008) reported DDT levels and egg shell thickness in three avocet eggs collected in 2004 from the Salton Sea area. The geometric mean DDE level was 1.14 ppm with a range of 0.83–2.1 ppm. Although shell thickness was measured, eggshell thinning was not assessed because no comparison measurements were available from eggs collected prior to the DDT era. According to Henny et al. (2008), Setmire et al. (1993) had concluded that the risk to American avocets for DDE in the Salton Sea area was not high in the 1988–1990 period. Black-necked stilt eggs collected in the same area as the American avocet eggs declined in DDE concentration from 2.91 to 0.55 ppm from 1993 to 2004. Based on this limited and indirect assessment, low ppm levels of DDE in American avocet eggs do not appear to be of reproductive concern among scientists studying the effects of DDT on avian wildlife.

DDT and Killdeer. No studies were found that related DDT levels in killdeer eggs to eggshell thinning or reproductive success.

Discussion and conclusions. There is a wide range of DDT levels seen in individual eggs collected from the same species and location. Authors have attempted to explain the wide variability in avian egg DDT levels. Important factors appear to be the age of the bird, level of contamination in feeding grounds, prey selection and nutritional status. Since avian wildlife accumulate DDT throughout their lives, older birds have higher body burdens and lay eggs with higher levels of DDT. Most species considered here do not reproduce until the third or fourth year of life. Therefore, dietary sources of DDT in the first 3–4 years of life will be the major source of DDT in eggs during the first year of reproduction. DDT in eggs thereafter will increase as a result of dietary DDT accumulated during each subsequent year. Egg laying and starvation can partially offset these life-long increases.

For Newport Bay, a hen laying eggs having DDT levels at the bottom of the range is probably in its first year of reproduction and has, by feeding location and prey selection, not been exposed to significant levels of DDT. These hens are most likely feeding several miles out in the Pacific Ocean in an area far from DDT contamination.

A hen laying eggs having DDT levels at the top of the range is probably an older bird (some live up to 20 years) with a history of frequenting locations and consuming prey that are high in DDT. In Southern California, the highest DDT levels in fish are found off of the Palos Verdes Peninsula. For several miles around the Los Angeles County Sanitation Districts outfall, the ocean bottom and aquatic biota have been contaminated by DDT manufacture wastes. The next highest levels are found in rivers draining from Mexico into California. The Alamo and New Rivers that drain into the Salton Sea and the Tijuana River that drains into the Pacific Ocean are examples. These rivers drain agricultural lands in Mexico where DDT was used long after the U.S. ban in 1972. In addition, the continuing use of DDT to control malaria along the southwestern coast of Mexico is also a significant exposure location. The next highest levels occur in rivers, estuaries and bays that receive agricultural drainage in California. Newport Bay is such an example, although for the Newport Bay Watershed the downward trend in DDT releases has been accelerated in recent decades by the conversion of agricultural lands to urban uses, a change that reduces the erosion of soils where DDT was historically used. The least contaminated aquatic biota can be found in areas of the Pacific Ocean not associated with contaminated effluents and in estuaries that do not receive drainage from lands containing DDT residues from past agricultural and other applications. Most of the fish-eating avian species resident to Newport Bay and Watershed are migratory, so all of the above exposure scenarios are possible.

When interpreting studies on the effects of DDT on avian reproduction, one must be aware of the wide range of egg residue levels. Average egg residue levels, average shell thinning and average hatching failure do not reveal the full range of effects (i.e., from no effect to severe effect) that can occur in individual birds at the same location and in the same data set. Thresholds should be estimated from the full range of effects in individual birds and not from the average effect for a particular study. For example, the average egg residue of 4 ppm DDT may result in a 30% hatching failure. Examination of the individual data may reveal that the hatching failure only occurred at egg levels above 10 ppm, so the threshold for hatching failure is closer to 10 ppm than 4 ppm.

Since the U.S. ban of DDT in 1972, residues in the aquatic environment and in bird eggs have declined to levels that, in most locations are below thresholds for eggshell thinning and in almost all locations, below thresholds for hatching failure.

The CH2M Hill study found that DDT levels in eggs of six avian species were below thresholds for hatching failure, with perhaps minimal egg shell thinning in one species. The lack of a correlation of DDT levels with shell thickness in Forster's tern suggests that the closely related and endangered least tern is not being affected by DDT. The IAP to the Organochlorine TMDL for Newport Bay and Watershed has recommended (IAP 2009) the least tern as a sensitive indicator for the potential toxicity of DDT to wildlife in Newport Bay.

3.1.6 DDT in Marine Mammals

In April 2006, and again in December 2013, a comprehensive review of the scientific literature was undertaken to assess what is currently known regarding the effects of DDT in marine mammals, either resident to, or capable of visiting, Newport Bay, California. The first step of the review was to determine the species that should be included. Although there are numerous marine mammal species found in the northwestern Pacific Ocean, relatively few species reside in, or visit, Newport Bay.

Those that may potentially reside in the area for significant periods include the California sea lion (*Zalophus californianus*) and harbor seal (*Phoca vitulina*). Those species that may enter Newport Bay for at least short periods—an unlikely but conservative approach—include the Pacific bottlenose dolphin (*Tursiops gilli*), rough-toothed dolphin (*Steno bredanensis*) and common dolphin (*Delphinus delphis*), and two filter-feeding baleen whale species—the minke whale (*Balaenoptera acutorostrata*) and the migratory gray whale (*Eschrichtius gibbosus*; Ingles 1965; Burt and Grossenheider 1976).

Therefore, electronic database searches were conducted via both the ISI Web of Science and BIOSIS Previews using the following topical keywords:

Seals and DDT
Sea Lions and DDT
Dolphins and DDT
Whales and DDT

Several hundred documents dating from the mid-1960s through 2013 were identified. However, most involved species not relevant to the Newport Bay region (i.e., not listed above). Notwithstanding, a significant number of reports were identified and are summarized below. Although no search can necessarily identify and locate all publications on a topic, those summarized below provide a reasonable summary of what is currently known regarding DDT in marine mammals that may either reside in or visit Newport Bay.

One important factor to consider in this review is the virtual lack of publications encountered that describe the toxic actions or endpoints of DDT in the subject marine mammals. There are two key reasons for this. First, logistically marine mammals are very difficult to directly utilize in the statistically-significant numbers needed for valid potency or other mechanistic investigations. Although sea otters may only weigh a few pounds, whales are excessively large and not practical to handle or house. Second, marine mammals have been protected by the United States Government for many years, which has significantly reduced access for any purpose, including research. Therefore, nearly all the papers published to date involve the measurement of DDT residues in tissues obtained from either live or dead (stranded and often decaying) animals. Such information can at least give an approximate estimate of the residues encountered by the subject marine mammals—and their ability to accumulate them. The following is a brief summary of the published reports involving DDT in marine mammals of importance to Newport Bay.

The summary is a chronology by species of DDT studies in marine mammals. DDT concentrations in these studies are reported as total DDT, which typically represents the sum of anywhere from three (p,p'-DDT+p,p'-DDD+p,p'-DDE) to six (o,p'-DDT+o,p'-DDD+o,p'-DDE+p,p'-DDT+p,p'-DDD+p,p'-DDE) analytes. Unless otherwise indicated, all residue values reported below are based on wet sample weight—concentrations reported on a lipid weight basis can average four or more times higher than those reported on a wet weight basis. Also note that while many reported values are geometric means (delineated below), some are arithmetic means.

DDT and Seals. DDT has been detected in harbor seals (*P. vitulina*) throughout the world for several decades. A series of early studies, centered on the North Sea coastline, document the DDT concentrations, with tissue type, commonly encountered when the insecticide was in widespread use (Koeman and van Genderen 1966; Koeman et al. 1972; Drescher et al. 1977; Duinker et al. 1979). DDT concentrations (in ppm) ranged as follows: blubber, 0.51–25.4; liver, 0.06–1.3; kidney, 0.05–0.76; brain, 0.038–3.1; spleen, 0.029–0.18; and heart, 0.25–0.60. It was obvious from an early date that fat-soluble DDT and its associated degradation products selectively partitioned into relatively inactive adipose tissue. Thus, while tissue-borne residues could be significant, the potential for toxic effects as a result would be both low and difficult to assess.

In response to declining harbor seal populations in the Dutch Wadden Sea (the southern coastal North Sea), Reijnders (1980) measured DDT concentrations (in ppm) in kidney, liver, and blubber (on a lipid weight basis) from resident harbor seals. In adult seals, mean DDT concentrations varied as follows: kidney, 0.2–0.9; liver, 0.4–2.1; and blubber, 8.5–47.3. He also determined that the decreased reproductive success reported for the Dutch Wadden Sea (vs. the German Wadden Sea) was strongly correlated to the tenfold higher PCB concentrations of the region; DDT was not strongly correlated with reproductive success.

In 1990, Luckas et al. reported mean DDT concentrations (in ppm) in harbor seals from a number of diverse geographic locations: Norway, 1.226; Sweden, 22.498; Iceland, 1.546; Germany, 3.903, and Antarctica, 0.105. Not surprisingly, higher concentrations were associated with regions of greater agricultural activity.

In 1992, Hall et al. compared DDT concentrations in both victims (34) and survivors (54) of a phocine distemper epizootic to determine if a correlation with the disease may exist, indicating a possible immunosuppressive role for DDT—one has been suspected for some chlorinated biphenyls. DDT concentrations ranged from 0.13 to 12.1 ppm for live animals and 0.71–7.17 ppm for dead animals; hence, no significant correlation could be made to indicate that DDT residues may have increased seal susceptibility to the disease.

Vetter et al. (1996) reported the mean DDT concentration for 32 harbor seals collected from the North Sea between 1988 and 1995 to be 3.903 ppm (range, 1.501–11.475). They also found no significant difference in the DDT concentrations between seal adults and pups collected prior to (1987) and during (1988) a major seal die-off, which indicated DDT was probably not the cause.

Routti et al. (2008) compared DDT levels in gray and ring seals in the more contaminated Baltic Sea compared to lower contaminated sites in Canada and Norway. They reported that changes in circulating vitamin D and thyroid hormone were associated with DDT and PCB levels in liver, suggesting that bone lesions observed in Baltic gray seals may be caused by DDT and/or PCBs. Another possible explanation for their findings is that animals stressed by factors unrelated to body burdens of DDT will result in weight loss with the mobilization of fat stores of DDT. The mobilization of DDT from blubber to liver could result in a negative correlation between DDT levels in the liver and the effects of the unknown factor, even though DDT may not be at a level that is having any toxic effect.

Bredhult et al. (2008) did not find an association between DDT in blubber and uterine leiomyomas in Baltic gray seals.

In 1997, Hayteas and Duffield reported the p,p'-DDE concentrations from the blubber of some ten harbor seals collected off the Oregon coast to have a geometric mean of 1.9 ppm (range, 0.4–12.5 ppm); p,p'-DDT levels were not reported as they were negligible in all samples. They concluded that DDT contamination along the Oregon coast was relatively low, and that animals with higher residue levels may have migrated from California. Moreover, in 1997, Mossner and Ballschmiter reported a mean DDT concentration from two harbor seals collected from the North Atlantic Ocean to be 18.99 ppm (on a lipid weight basis).

More recently, Kajiwara et al. (2001) reported DDT concentrations (based on lipid weight) in the livers of ten stranded harbor seals collected between 1991 and 1997; the geometric mean concentration was 12 ppm (range, 2.8–85 ppm).

Greig et al. (2011) reported DDT levels (lipid basis) of 320–1,500,000 ppb in blubber from 202 stranded and wild-caught harbor seals in the central California coast. The highest levels were in pups during the post-weaning fast, suggesting that fasting pups may be the most vulnerable early life stage to the toxic effects of DDT, due to mobilization from lipid stores.

Hall et al. (2009) reported a negative correlation between blubber concentrations of DDT and survival in first year gray seal pups sampled from the Isle of May in 2002. Geometric mean DDT levels (lipid basis) were 229 ppb. The authors were careful to note that their finding did not indicate causation. Causation would seem unlikely considering the relatively low concentrations of DDT in the pups.

Fillman et al. (2007) reported 20–2,480 ppb DDT (mean level of 660 ppb) in blubber from stranded juvenile South American fur seals. The authors point out that the poor nutritional status of the seals is likely to have mobilized DDT from blubber to other tissues, increasing their vulnerability to possible toxic effects.

Roos et al. (2012) reported a decline in DDT residues in seal blubber from 192 to 2.8 ppm (lipid weight basis) from 1973 to 2010. Increases in uterine health and pregnancies and reduced uterine cancer were associated during the same period with declining residues of persistent organochlorines, including DDT, suggesting possible causation.

In recent years, DDT contamination of harbor seals in the U.S. was re-evaluated because DDT's ban has been in place for well over 30 years. Shaw et al. (2005) sampled the blubber of 30 stranded harbor seals from the northwestern Atlantic coast of the U.S. DDT concentrations ranged from 1.4 to 57.5 ppm (lipid weight). Also of note was substantial variation between adult males (12.40 ± 6.65 ppm), adult females (4.60 ± 2.56 ppm), yearlings (13.00 ± 14.40 ppm), pups (21.10 ± 19.70 ppm), and fetuses (2.21 ± 0.62 ppm).

Wang et al. (2007) reported relatively low levels of DDT in harbor seals from the Gulf of Alaska. Blubber samples contained 78–325 ppb with a mean level of 159 ppb. The authors point out that levels in nursing females were much lower than those in male adults, due to lactation transfer of DDT from mother to newborns.

Sakai et al. (2006) reported on an assay for androstane receptor activity in Russian seals. The authors concluded that DDT was active in their assay and that the lowest observable effect level was comparable to a 10 ppm tissue level.

In summary, to date, a number of investigations have confirmed the presence of DDT in harbor seals throughout the world, and their ability to accumulate it via primarily biomagnification. Concentrations vary but have been generally reported in the parts-per-million range, which would reflect the varied length of use of the insecticide (although banned in 1972 in the U.S., it was used much more recently in other parts of the world), as well as their habit of feeding high on the marine food web (primarily fishes). Toxic effects in harbor seals from DDT have yet to be conclusively demonstrated via controlled studies.

DDT and Sea Lions. There are several reports of DDT in sea lions (*Z. californianus*) residing along the California coast. In 1971, Le Boeuf and Bonnell published a seminal report of blubber concentrations in California sea lions collected in 1970 (n = 12), a full 2 years prior to the banning of the use of DDT in the U.S. In it, they reported high concentrations for both DDT (geometric mean, 17 ppm; range, 8.8–34 ppm) and DDE (geometric mean, 740 ppm; range, 370–1,500 ppm).

In 1992, Bacon et al. surveyed milk samples from a number of pinniped species, including one lactating California sea lion resident to the central coast—geometric mean values ranged from 3.3 ppb for o,p'-DDT to 1.4 ppm for p,p'-DDE. This was not considered unusual, as the area is one of intense agricultural activity and has a history of DDT use.

In 1995, Lieberg-Clark et al. followed up on the above 1971 report of Le Boeuf and Bonnell by measuring ΣDDT and ΣDDE concentrations in blubber from seven California sea lions sampled between 1988 and 1992. Their numbers clearly indicated a significant decline (greater than 99%) in residues over the 20-year time span for both DDT (geometric mean, 0.16 ppm; range, 0.07–0.35 ppm) and DDE (geometric mean, 5.0 ppm; range, 2.5–10 ppm). Therefore, they concluded the following:

1. The decline in the residue levels in California sea lions over this period was accompanied by a significant increase in their population during the same time period.

2. The extremely high ΣDDT concentrations reported in the 1970s may have been associated with reproductive problems in California Sea Lions.
3. The decline in ΣDDT residues in California sea lions was so dramatic because their breeding area in southern California was much less contaminated with DDT residues than in 1970.

However, O'Shea and Brownell (1996) took issue with the latter statement, which they considered to be based primarily upon circumstantial evidence. For instance, they suggested that the original sample sizes (7 and 12) were too limited to draw such sweeping conclusions. In addition, they noted a paucity of experimental evidence demonstrating an impact of DDT and/or it metabolites on sea lion reproduction. In addition, O'Shea and Brownell (1996) noted that California sea lion populations have historically fluctuated, declining in the late 1800s and early 1900s, and increasing in the 1960s. Therefore, while they do not necessarily discount the observations of Lieberg-Clark et al. (1995), their overall contention was that to-date there was insufficient evidence to draw such conclusions.

In 1997, Hayteas and Duffield reported the p,p'-DDE concentrations from the blubber of some five California sea lions (in addition to harbor seals, above) collected off the Oregon coast to have a geometric mean of 8.1 ppm (range, 3.2–15.4 ppm); p,p'-DDT levels were again not reported as they were negligible in all samples. They again concluded that animals with higher residue levels may have migrated from California. In addition, and most importantly, their p,p'-DDE value was in the same range as that of the Lieberg-Clark et al. (1995) study, providing further confirmation of the dramatic decline in residues reported by them.

More recently, Kajiwara et al. (2001) reported the concentrations of organochlorine insecticides (based on lipid weight) in some 15 stranded California sea lions collected between 1991 and 1997; in blubber, the geometric mean DDT concentration was 209 ppm (range, 13–2,900 ppm), while in liver it averaged 142 ppm (range, 12–970 ppm). Their results contrasted with those of Lieberg-Clark et al. (1995) for animals collected during an overlapping time period; however, the Lieberg-Clark et al. (1995) data were reported on a wet sample weight basis.

Connolly and Glaser (2002) reported the accumulation of p,p'-DDE in female California sea lions resident to the California Channel Islands. High concentrations of DDT and its degradation products emanating from the Whites Point outfall contaminated the sediments and aquatic life of the Palos Verdes shelf and Santa Monica Bay. Fish contaminated by these DDT wastes were suspected of serving as vectors in transferring residues to the sea lion population. However, they determined that p,p'-DDE residues in the blubber of female premature parturient sea lions from San Miguel Island declined from a mean of 944 ppm in 1970 to 40 in 1991, while those from full-term parturient females also declined during the same time period (from 109 to 10 ppm). Both declines, approximately a full order of magnitude, were similar to that reported by Lieberg-Clark et al. (1995) and mirror the declines observed in sediments and mussels. In addition, Connolly and Glaser (2002) noted that reduced concentrations in full-term parturient females were most likely influenced by lactation.

As follow up to the 1971 study, Le Boeuf et al. (2002) revisited the topic of organochlorine pesticides in marine mammals. They collected blubber samples from some 36 stranded animals along the coast of California in 2000, and determined geometric mean DDT concentrations of 37 ± 27 ppm (wet weight basis) and 150 ± 257 ppm (lipid weight basis). They found no significant differences in concentrations with differences in age or sex, but did conclude that DDT levels decreased by more than one order of magnitude between 1970 and 2000. Kannan et al. (2004) also reported the results of DDT analysis performed on the blubber of some 36 stranded California sea lions collected in 2000. As Kannan is a co-author of the Le Boeuf et al. (2002) study, it is unclear if the animals used were the same in both studies. However, he reports a mean DDT concentration of 143 ± 253 ppm, with a geometric mean of 69 ppm.

More recently, two studies designed to correlate toxic actions with DDT in California sea lions have been published. Debier et al. (2005) investigated a possible relationship between DDT in the serum of 12 healthy California sea lions and circulating levels of vitamins A and E and the thyroid hormones thyroxine (T4) and tri-iodothyronine (T3). Although several negative correlations were reported for PCB, only vitamin A was significantly correlated with DDT, and only when concentrations were reported on a lipid weight basis.

In 2005, Ylitalo et al. used a logistic regression model with California sea lions in an attempt to correlate the unusually high prevalence of neoplasms (carcinomas—found in 18% of stranded adults) with blubber DDT concentrations. Although concentrations were significantly higher in animals that died from carcinomas versus those that did not, after controlling for other confounding factors only blubber thickness proved to be a reliable predictor of death via carcinoma—ultimately DDT was proven not significant.

Blasius and Goodmanlowe (2008) reported that DDT levels in blubber collected from marine mammals in the southern California bight were higher in resident harbor seals and sea lions than in the transient northern elephant seal. Adult female sea lions had lower residue levels than pups, yearlings and adult males. DDT levels in sea lion blubber declined approximately tenfold to a mean (lipid weight basis) of approximately 200 ppm from 1994 to 2006, but not in the transient northern elephant seal that was less impacted by the high levels of contamination attributed to production wastes released prior to 1970. The highest concentrations of DDT in blubber, as high as 13,271 ppm (lipid weight basis), were measured in stressed sea lions that had lost almost all of their blubber.

In contrast to the high blubber residues of DDT in the highly contaminated southern California bight, blubber from sea lions stranded along the Baja California coast in 2000 and 2001 had residues averaging 3.8 ppm on a lipid weight basis (Del Toro et al. 2006). Nino-Torres et al. (2009) reported a mean DDT level of 3.4 ppm (lipid weight basis) in blubber collected from sea lions in 2005 and 2006 in the Gulf of California.

Ramsdell (2010) reported a novel zebra fish model for the interaction of DDT and the diatom poison, domoic acid, in sea lions feeding in the highly contaminated Channel Islands of the southern California bight. Pretreatment of embryonic zebra fish with levels of DDT comparable to those found in more highly contaminated sea lion fetuses, increased the neurological response to domoic acid. This model suggests that high residues of DDT, typically found in sea lions residing near the Channel Islands may increase susceptibility to domoic acid poisoning.

In summary, several studies have confirmed the presence of DDT in California sea lions. Hence, they accumulate DDT primarily via biomagnification by feeding on contaminated fish, as do seals. DDT concentrations have generally been reported in the parts-per-million range, but have been on the decline in recent years from discontinuation of its use. The highest DDT level in sea lions world-wide occurred in the Channel Island area of the southern California bight as a result of past dumping of DDT manufacture wastes there. Similar to harbor seals, toxic effects from DDT in California sea lions have yet to be conclusively demonstrated.

DDT and Dolphins. Dolphins and porpoises are not likely to spend much time (if any) in Newport Bay. However, to be conservative they have been included in this review. There are relatively few published reports of DDT in dolphins and porpoises that might be relevant to Newport Bay. In 1980, O'Shea et al. reported DDT in the blubber, brain and muscle tissues of 69 small cetaceans, including one Pacific bottlenose dolphin (*T. gilli*) that had an excessively high blubber DDT concentration of 2,695 ppm.

Smyth et al. (2000) reported concentration ranges of DDT in the blubber and liver of six common dolphins (*D. delphis*) accidentally caught in fishing nets off the coast of Ireland to range from 3,998 to 9,444 ppb and 2,293 to 4,528 ppb, respectively. In 2001, Borrell et al. reported the DDT concentrations measured in the blubber of common dolphins accidentally caught in fishing nets along both the Atlantic and Mediterranean coasts of Spain during a 12-year time span. In dolphins from the Atlantic, mean DDT concentrations (lipid weight basis) declined significantly between 1984 and 1996 (1984: 15.54±8.82 ppm; 1996: 7.95±4.49 ppm). In dolphins from the Mediterranean, mean DDT concentrations of animals sampled in 1992 through 1994 was 33.40±38.64. Of note was the fact that males in both regions accumulated significantly higher concentrations than females. In a follow-up study, Borrell and Aguilar (2007) reported again on DDT levels in dolphins from Spain's Mediterranean coast. Levels continued to decrease in bottlenose dolphins, a reduction from 303 to 13 ppm lipid weight basis from 1978 to 2002. The ratio of p,p'-DDE to p,p'-DDT continued to increase, suggesting the continued breakdown of DDT with the absence of new releases into the environment.

Castrillon et al. (2010) reported a sixfold decrease (from approximately 400–70 ppm lipid weight basis) in blubber DDT in striped dolphins (*Stenella coeruleoalba*) dying from two Mediterranean epizootics, due to morbillivirus in 1990 and 2007. The second epizootic was not believed to have been enhanced by DDT or PCBs residues in the dolphins. Wafo et al. (2012) reported an average DDT level (lipid weight basis) of 16 ppm in blubber collected in 2007–2009 from striped dolphins stranded along the French Mediterranean coast. Shoham-Frider et al. (2009) reported DDT levels of 0.92–141 ppm in blubber from bottlenose dolphins stranded in 2004–2006 along the Israeli coast.

Das et al. (2006) pointed out that even though DDT levels in blubber correlated with thyroid fibrosis in harbor porpoises from northern Europe, this type of association was insufficient to establish a cause-effect relationship. DDT levels (lipid weight basis) were highest (viz., 1,481–2,292 ppb) in the more industrialized areas and were lowest at 1,122 ppb in the less industrialized Icelandic coast.

Siebert et al. (2011) found no correlation between stress hormones and DDT levels in blood collected in 1997–2002 from free-ranging and captive harbor porpoises (*Phocoena phocoena*) from the North and Baltic Seas. Geometric mean DDT levels in blood ranged from 0.2 to 8.2 ppm on a lipid weight basis. Weijs et al. (2010) reported a decline in DDT in tissues of harbor porpoises in the southern North Sea during the period 1990–2008.

Fair et al. (2010) reported levels of DDT (lipid weight basis) in blubber biopsies taken from bottlenose dolphins in 2003–2005 along the southeastern US coast. Adult males were highest at a geometric mean level of 29 ppm for Charleston, South Carolina. At the same location, adult females were lowest at 3.0 ppm with juvenile dolphins intermediate at 14.7 ppm. Litz et al. (2007) measured DDT in dolphin blubber biopsies taken in Biscayne Bay, Florida in 2002–2004. The geometric mean level in adult males and juveniles was 2.98 ppm, whereas the geometric mean level in adult females was lowest at 0.097 ppm.

Lailson-Brito et al. (2011) reported DDT levels (lipid weight basis) of 264–5,811 ppb in blubber of franciscana dolphins stranded along the Brazilian coast. In a follow-up study, Lailson-Brito et al. (2012) measured DDT levels in blubber from four species of dolphins. Much higher p,p'-DDT/p,p'-DDE ratios were measured in blubber from Fraser's dolphins that feed in deep open water, where slower breakdown of p,p'-DDT to p,p'-DDE would be expected. The blubber of Fraser's dolphins had relatively low residue levels of 0.99 ppm DDT. Higher levels of DDT at 5.0 ppm, 26.4 ppm and 2.4 ppm were measured in bottlenose, rough-toothed and long-beaked common dolphins, respectively, that inhabit estuarine and coastal waters that are more contaminated from DDT agricultural use.

Law et al. (2013) reported DDT levels in blubber from 43 female short-beaked common dolphins (*Delphinus delphis*) bycaught in fisheries during the period 1992–2006 off the southwest coast of the UK. Levels ranged from 0.2 to 16.1 ppm on a lipid weight basis. DDT levels declined during the period of investigation although the trend was not statistically significant.

Stockin et al. (2007) reported DDT levels in blubber from common dolphins in Hauraki Gulf, New Zealand from 1999 to 2005. Levels ranged from 17 to 337 ppb in females and 654–4,430 in males. In follow-up studies, Stockin et al. (2010) reported on levels of DDT in Hector's dolphin (*Cephalorhynchus hectori hectori*) and Maui's dolphin (*Cephalorhynchus hectori maui*) from 1997 to 2009 from all parts of New Zealand. DDT in blubber of females ranged from 94 to 8,210 ppb with a mean of 1,358 ppb, whereas DDT in male blubber ranged from 252 to 57,390 ppb with a mean of 12,400 ppb. DDT residues in individual dolphins reflected the proximity of their habitat to agricultural and industrial releases of DDT.

Wu et al. (2013) reported levels of DDT ranging from 0.845 to 179 ppm (with a mean of 64.2 ppm) in blubber from Indo-Pacific humpback dolphins (*Sousa chinensis*) stranded in 2004–2009 in the Pearl River Delta in China. Seventy five percent of the DDT residue was p,p'-DDT, suggesting fresh releases of DDT. The authors note that DDT is still being used in China for control of malaria.

In summary, there are few reports of DDT concentrations in dolphins or porpoises important to the Newport Bay region. Those above are for animals sampled elsewhere in the world. Although they demonstrate the ability of several species of dolphins to accumulate DDT and its degradation products, the actual concentrations may not reflect what would occur in animals residing on the California coast. Similar to harbor seals and California sea lions, toxic effects from DDT in the subject dolphins have yet to be conclusively demonstrated via controlled studies.

DDT and Whales. Although whales (baleen or toothed) are not likely to spend time in Newport Bay, to be conservative, we summarize the pertinent publications involving DDT and the whale species most likely to, at least, briefly visit the area. Over the years, many studies have reported on the contaminants present in the blubber of baleen whales, including gray and minke whales. For instance, in gray whales (*E. gibbosus*) Wolman and Wilson (1970) measured DDT concentrations as high as 680 ppb in some 23 animals collected between 1968 and 1969, and Schafer et al. (1984) reported a concentration of 470 ppb in a single animal sampled in 1976. In 1994, Varanasi et al. reported the concentrations of DDE in the tissues and stomach contents from 22 gray whales stranded between 1988 and 1991 along the coast from Kodiak Island, Alaska, to San Francisco, California. Gray whales have the unique habit of filter feeding along benthic sediments. Therefore, they are potentially capable of ingesting sediment-sorbed organic contaminants. Mean concentrations, and the ranges, measured in blubber were: DDT, 68 ± 22 ppb (1–370 ppb); DDD, 76 ± 24 ppb (1–470 ppb); and DDE, 310 ± 96 ppb (9–2,100 ppb). In liver, residues were predictably reduced: DDT, 1 ± 0.4 ppb (0.4–3 ppb); DDD, 23 ± 5 ppb (0.6–52 ppb); and DDE, 100 ± 28 ppb (7–280 ppb). Most interestingly, they found no significant differences in the concentrations from whales collected in the more pristine Kodiak Island/Washington outer coastal areas versus those collected in the more impacted areas of Puget Sound, Washington, and San Francisco.

Tilbury et al. (2002) sampled gray whales from a subsistence harvest in the Arctic during the fall of 1994 and compared their DDT concentrations (per lipid weight) with those of stranded gray whales from the same general collection area. They discovered significant differences in the harvested versus stranded whale blubber concentrations of males (200 ± 38 ppb versus $39,000 \pm 23,000$ ppb), females (360 ± 66 ppb versus $2,800 \pm 1,000$ ppb) and juveniles (330 ± 53 ppb versus $11,000 \pm 4,300$ ppb), respectively. The consistently higher concentrations in stranded animals may indicate their possible cause of death. However, tissue degradation of dead and potentially decaying animals limits the usefulness of such a comparison.

In minke whales, Schafer et al. (1984) reported a DDT concentration of 587 ppm from a single animal stranded off southern California. However, this high concentration appears to be linked to an urbanized area, as 29 minke whales sampled off the South African coast ranged only as high as 820 ppb (Henry and Best 1983), while another 37 sampled in Antarctica ranged from 10 to 140 ppb (Tanabe et al. 1986).

In 1998, Klevaine and Skaare published their findings on the chemical concentrations in some 72 minke whales stranded along the northeastern Atlantic seaboard (coastal Norway, West Spitsbergen Island, and Bear Island) in 1992. Although they found no significant differences in mean DDT concentrations between juvenile males versus females (1.94 ppm versus 2.77 ppm lipid weight, respectively), they did conclude differences existed between adult males and females (3.86 ppm versus 1.51 ppm, respectively), as well as between juveniles and adults (both males and females).

The DDT concentrations were also determined for some 155 minke whales harvested in 1998 from the North Atlantic and European Arctic Oceans (Hobbs et al. 2003). Results ranged from 65.3 to 6,280 ppb (lipid weight basis), which encompass the concentrations measured in whales taken 6 years earlier by Klevaine and Skaare (1998).

Finally, in one of the few mechanistically-oriented papers involving any cetacean, Niimi et al. (2005) reported the full-length cDNA sequences of two cytochrome P450 (CYP) isozymes, from minke whale liver, responsible for either the bioactivation or detoxification of xenobiotic chemicals. While CYP1A1 consisted of 516 amino acid residues and was deemed most closely related to that from sheep and pigs, CYP1A2, also consisting of 516 residues, was deemed most closely analogous to that from humans, indicating that the enzyme's function in minke whales may be similar to that of humans. However, Niimi et al. (2005) found no significant correlation between hepatic DDT levels and mRNA expression levels of CYP1A1 and CYP1A, indicating that DDT may not be responsible for their induction in minke whales.

Lailson-Brito et al. (2012) reported DDT levels (lipid weight basis) of 125.6 ppm in blubber from a female killer whale (*Orcinus orca*) and 17.9 ppm in a neonatal female false killer whale (*Pseudorca crassidens*) from the southern Brazilian coast. Considering that female marine mammals tend to have lower residues of DDT than males due to lactation, male killer whales would have even higher residues than existed in the females mentioned above. Transient killer whales, known to feed primarily on marine mammals (the authors of this study reported the presence of a dolphin remains in the stomach of the killer whale), are at the highest point in the marine food-chain and therefore would be expected to have the highest residues of DDT as reported in this study.

Elfes et al. (2010) collected blubber biopsies from adult male humpback whales (*Megaptera novaeangliae*) along the North American Pacific coast in 2003–2004 and in the Gulf of Maine in 2005–2006. Whales sampled in southern California had the highest levels (lipid weight basis) of DDT at 4,900 ppb. Residues decreased more than one order of magnitude as the sampling points moved north to the Bering Sea. DDT residues increased with age.

Hoguet et al. (2013) sampled blubber in Beluga whales (*Delphinapterus leucas*) from the Alaskan artic from 1989 to 2006. The median DDT level (lipid weight basis) in blubber was 1,890 ppb. Male residues were more than three times the residue level in females. The authors note that the wide fluctuation in seasonal blubber thickness may result in mobilization of DDT from lipid stores in blubber.

McKinney et al. (2006) reported liver DDT levels (lipid weight basis) in beluga whales from the Saint Lawrence Estuary and Hudson Bay. The average residue level of DDT was 4,536 ppb for whales from the Saint Lawrence Estuary and 284 ppb for whales from Hudson Bay.

Nino-Torres et al. (2010) reported in DDT levels in blubber biopsies from fin whales (*Balaenoptera physalus*) collected in 2004–2005 in the Gulf of California, Mexico. DDT levels ranged from 200 to 1,900 ppb with an average of 500 ppb in females and 850 ppb in males. These levels are among the lowest reported worldwide for fin whales and other marine mammals.

In summary, the few studies reporting DDT residues in gray and minke whales indicate that they are also capable of accumulating residues in their blubber and other tissues. However, since they feed fairly low on the marine food web (invertebrates), their residue levels tend to be relatively lower than those of fish-eating marine mammals (viz., seals, sea lions, and dolphins). Similar to the other marine mammals discussed above, toxic effects from DDT in gray and minke whales have yet to be conclusively demonstrated.

Conclusions. The above DDT residue data in marine mammals indicates that fat-soluble DDT and/or its degradation products have been detected in many marine mammalian species worldwide since the mid-1960s. In general, during the DDT-use era blubber concentrations in the parts-per-million range were not uncommon, particularly for the species that feed primarily on fishes—thus higher on the marine food web. Of importance to the Newport Bay region are the harbor seal, California sea lion, and the Pacific bottlenose and common dolphins. Clearly, marine mammals are quite capable of accumulating residues as long as DDT continues to exist and accumulate in the aquatic food chain. However, in the years since the U.S. ban on DDT use, tissue concentrations have decreased in tandem with the decline in environmental concentrations. A similar trend has been observed for gray and minke whales. However, since baleen whales tend to feed at lower levels of the marine food web, blubber concentrations have tended to be an order of magnitude lower—in the parts-per-billion range.

In this literature review we also sought to assess the role of DDT in causing possible embryo deformities and/or other measurable health effects in marine mammals. However, marine mammals are a unique class or animals, in that published reports on controlled studies documenting such toxic effects were not encountered. There appear to be two reasons for this. First, they are too large and heavy to be easily housed, handled and utilized in controlled experiments with sample sizes sufficient to provide for statistical analysis. Second, they have been strictly protected by the federal government for many years, which has severely limited access. As a result, and as can be deduced from the chronology above, nearly all studies involving marine mammals and toxicants have been limited to residue analyses involving either dead/decaying animals or live, captive animals sampled via blubber biopsies.

Such a restriction has limited the field to speculation of effects, based upon measured residues and, in some cases, weak correlations. However, since metabolic activity in blubber is relatively low, it is assumed that large concentrations would need to be attained before measurable effects would be observed. Thus, to date, few if any toxic impacts have been clearly delineated for DDT in the marine mammals that are the focus of this report, and with tissue residues clearly on the decline, the likelihood that they might be identified in the future is declining.

3.2 TMDL Targets

The Regional Board's TMDL targets for DDT in the Newport Bay and Watershed (SARWQCB 2006) are summarized in Table 5. In addition, wildlife guidance for a fish diet by Environment Canada (2000) and by US EPA (1995) has been considered as potential TMDL targets for Newport Bay and Watershed. The following sections address the science underlying each of the proposed and considered targets.

3.2.1 Sediment TELs

The decision by US EPA Region IX (2002) and the Regional Board (SARWQCB 2006) to use threshold effect levels (TELs) as TMDL sediment targets raised concern among the regulated community, because TELs do not fully consider dose-response. TELs rely to a considerable extent on the occurrence of a chemical in toxic sediments

Table 5 Proposed sediment, fish, and water column TMDL targets for total DDT in the Newport Bay Watershed

Sediment targets (µg/kg dry wt)[a]	
San Diego Creek and tributaries	6.98
Upper and Lower Newport Bay	3.89
Fish tissue targets for protection of human health (µg/kg wet wt)[b]	
San Diego Creek and tributaries	100
Upper and Lower Newport Bay	100
Fish tissue targets for protection of wildlife (µg/kg wet wt)[c]	
San Diego Creek and tributaries	1,000
Upper and Lower Newport Bay	50
Water column targets (µg/L)[d]	
San Diego Creek and tributaries	
Acute Criterion (*CMC*)	1.1
Chronic Criterion (*CCC*)	0.001
Human Health Criterion	0.00059
Upper and Lower Newport Bay	
Acute Criterion (*CMC*)	0.13
Chronic Criterion (*CCC*)	0.001
Human Health Criterion	0.00059

[a]Buchman (1999)
[b]Pollock et al. (1991); Brodberg and Pollock (1999)
[c]National Academy of Sciences (NAS 1972)
[d]US EPA Region IX (2000)

rather than on direct causation. In addition, TELs for total DDT are many orders of magnitude below thresholds for DDT in spiked sediment bioassays. These scientific issues triggered a more detailed analysis of the freshwater and marine TELs for total DDT.

The TEL sediment targets for total DDT were those summarized by Buchman (1999) from the original publications by MacDonald et al. (1996) and by Smith et al. (1996). As shown in Table 5, the targets are 6.98 ppb for the fresh water drainage into Newport Bay and 3.89 ppb for the estuarine and salt water of the Bay.

Several key authors in the TMDL regulation for Newport Bay and Watershed have expressed reservations about TELs for total DDT. Buchman (1999), author of the National Oceanic and Atmospheric Administration (NOAA) table listing the sediment targets, wrote:

> These tables are intended for preliminary screening purposes only: they do not represent official NOAA policy and do not constitute criteria or clean-up levels.

MacDonald et al. (MacDonald et al. 1996; Smith et al. 1996) authors of the primary references cited by Buchman wrote:

> Low reliability (TS=0) was indicated for only one substance (total DDT).

Peter Kozelka and David Smith at US EPA Region IX, authors of the 2002 TMDL report for Newport Bay and Watershed, wrote:

> We recognize these NOAA values have been derived by associating nationwide sediment chemistry data sets with benthic toxicity results and there is no direct cause and effect relationship.

The Regional Board's organochlorine TMDL technical report (SARWQCB 2006), addressed sediment quality guidelines (SQGs), which include TELs as one kind of SQG:

> SQGs should be used with caution since individual SQGs are often unreliable indicators of toxicity and do not necessarily identify the correct cause of toxicity (Vidal and Bay 2005). In particular, use of empirically-derived marine SQGs for DDT and PCBs has been found to be relatively inaccurate in predicting toxicity (Long et al. 1995).

A detailed analysis of the TEL method and the data sets used to derive the TELs for total DDT has revealed much more of concern that should be considered and resolved before TELs are used as TMDL targets.

To gain a full understanding of the individual data points from which the marine and fresh water sediment TELs for total DDT were derived, each data point in the TELs reported by Buchman (1999) as cited by US EPA Region IX (2002) was reviewed. The data points and reference citations were purchased from the author of the TELs.[5] The data sets are slightly different from the ones used to derive the published TELs.[6] The derivation of the TELs is explained by MacDonald et al. (1996) as follows:

> "For each analyte, a TEL was derived by calculating the geometric mean of the 15th percentile of the effects data set and the 50th percentile of the no effects data set."

---

[5] MacDonald DD, BEDS data bases for total DDT purchased in 2005.
[6] Personal communication from DD MacDonald to J. Byard (2005).

Table 6 The no-effects and effects data sets for total DDT in marine and fresh water sediments

Total DDT in marine sediments[a]		Total DDT in fresh water sediments[a]	
No-effects data set	Effects data set	No-effects data set	Effects data set
0.8	0.4	0.384	1.5
1.04	0.7	0.384	1.9
1.08	1.5	0.384	5
1.27	1.58	0.384	7
1.39	1.6	0.384	9
1.5	2.92	0.384	10
3.42	2.92	5	10
4.6	2.93	5	50
4.94	3.27	5	120
5	15.4	5	222
5	18.8	5.42	4,200 ± 125
5	22.3	6	4,800
5	24	19.6 ± 18.4	11,000 ± 650
5	24.2	50.7 ± 119	11,100 ± 190
5	27	200	16,100
5	27	1,300	16,100
5	27	1,800	19,600 ± 2,180
5	29.7	5,300	21,100
5.18	33.2	5,800	30,600
5.23	46.1	22,100	46,300
5.38	54.5		49,600
6.9	55.2		49,700 ± 3,030
8.38	69		71,300
11	125		88,400
11.6	210		198,000
14.1	350		
22.5	432		
50	505		
100	596		
1,018	665		
2,170	3,000		
35,300	13,420		
	14,190		
	16,500		
	18,260		

[a]All values are in ppb dry weight in ascending order
Shaded values represent the 50th percentile of the no-effects data sets and the 15th percentile of the effects data sets
Data from DD MacDonald (personal communication)

The no-effects data sets and the effects data sets for marine and fresh water sediments are listed in Table 6 above.

Inspection of the four data sets raises the important question of why there are no effects at thousands of ppb and effects at a few or even less than 1 ppb. As will become apparent, these discrepancies are not so much the result of species and ecosystem differences, but rather, errors and misinterpretations in the data sets.

Using the 50th percentile of the no-effects data set ignores the most important information. The highest half of the data points that were not associated with a toxic effect are counted only by number and not by concentration. For example, the 50th percentile of the no-effects data set for total DDT in marine sediments is 5 ppb (average of two shaded values in Table 6). However, in the same data set, levels from 5 to 35,300 ppb are also without effect, but do not weigh by concentration of total DDT into the TEL value. Based on this data set, the threshold for toxicity is likely to be three orders of magnitude higher than the 50th percentile value. The threshold for toxicity is not appropriately weighed by the TEL calculation. This point has also been made by MacDonald, the author of the TELs. When MacDonald assessed total DDT toxicity in marine sediments from Southern California (MacDonald 1994), he reported a lowest observed effect level for total DDT for the most sensitive life stage of the most sensitive organism to be 7,120 ppb.

The 15th percentile of the effects data set for marine sediments is 2.92 ppb total DDT (shaded data point in Table 6 above). This means that only six out of 37 data points influence the 15th percentile by concentration. Twenty two data points matter only by count. Hence, whatever dose-response data are in the data set, only the 15th percentile value weighs into the TEL. Review of the underlying studies for each data point in the effects data set for marine sediments revealed highly significant errors and inconsistencies. The analysis is summarized below.

The data point of 1.6 ppb (Lyman et al. 1987), one of the 6 data points that count in the 15th percentile, is a re-publication of the 1.58 ppb data point (JRB Associates 1984) shown in Table 6. Hence, a single data point, based on equilibrium partitioning, is given the weight of two data points (Table 6).

The lowest five data points all rely on outdated and inaccurate K_{ow} values. K_{ow} values were used to estimate K_{oc} values that were then employed to determine equilibrium partitioning between water and organic carbon in sediment. The old methodology of determining K_{ow} has been superseded by the slow-stir method (de Bruijn et al. 1989). There is general consensus among scientists measuring K_{ow} that the slow stir method gives the most accurate value, and deserves to supersede earlier methodologies. K_{ow} values based on the slow stir method are recommended by U.S. EPA (2002) and are listed for the DDTs in the SARWQCB staff TMDL report (SARWQB 2006). The five data points relying on outdated K_{ow} values display errors of more than one order of magnitude.

Dealing with mixtures of chemicals becomes very important when evaluating many of the TEL data points in the total DDT effects data sets, particularly the five data points ranging from 2.92 to 3.27 ppb in marine sediments from highly polluted areas in San Francisco Bay (Table 6). Based on spiked-sediment bioassays and equilibrium partition calculations, these levels are three orders of magnitude below effect levels for total DDT. Yet, they are included in the effects data set and the lowest value is within the 15th percentile. Inspection of the underlying study (Chapman et al. 1987) revealed pollutant levels of 9,460 ppb PAHs (polyaromatic hydrocarbons), 129,000 ppb Pb, 710 ppb Hg, and 990 ppb Cu. Here, the problem is one of toxicological plausibility. The plausible explanation for benthic toxicity is the presence of these other pollutants and not the presence of low ppb levels of total DDT. The TEL method ignores obvious toxicological interpretation of data.

The effects data points of 22.3, 46.1, 54.5, 55.2, and 125 ppb are all LC_{50} values in pore water (Word et al. 1987), rather than LC_{50} values in sediments. These values should be multiplied by the partition coefficient (a factor of at least tens of thousands) to obtain estimates of sediment LC_{50s}.

The first of the two 27 ppb AET (apparent effects thresholds) values in the marine sediment effects data set is the highest threshold associated with *Rhepoxynius abronius* toxicity for total DDT in sediments from Northern California (Becker et al. 1989). The highest threshold represents the highest concentration found in sediments without toxicity. The comparable value for Southern California, where sediments have much higher levels of total DDT, due to contamination from a DDT manufacturing site, is greater than 9,300 ppb! The second 27 ppb value is the highest threshold associated with bivalve toxicity for total DDT in sediments from Northern California (Becker et al. 1989). A similar threshold was not determined for Southern California. The AET values for Northern California appear to be artifacts of the method (most likely determined by the presence of toxic levels of other contaminants), since sediments from Southern California with much higher residues of total DDT were not toxic in the selected bioassays.

The association between sediment residue and sediment toxicity at 68 ppb makes no sense when one considers that in the same study 1,018 ppb total DDT in sediment was not associated with significant sediment toxicity to the same amphipod species (Anderson et al. 1988). The authors stated: "Most notably, DDT concentration did not correlate with short-term toxicity or macrofaunal patterns."

The value of 210 ppb (Lyman et al. 1987) is derived in the same way as the 1.58 ppb value by JRB Associates (1984). The only difference is the use of the National acute marine criterion of 130 pptr instead of the chronic marine criterion of 1 pptr in the water column. The sediment equilibrium concentration is derived from a log K_{ow} of 5.98. The slow-stir log K_{ow} reported by de Bruijn et al. (1989) is 6.914. The K_{ow} derived by the superior slow-stir method gives a sediment acute marine threshold nearly an order of magnitude higher.

Using a method similar to the one used to estimate the 505 ppb data point, Neff et al. (1986) derived a screening level concentration for fresh water of 1.9 ppb. How can fresh and salt water screening levels differ by 265-fold when the toxicity of DDT to fresh and marine benthic organisms is similar? One or both of the screening levels are most likely in error. Based on bioassay results, the freshwater screening level appears to be too low.

The claim that DDT is toxic to benthic organisms at low ppb levels in marine sediments is not supported by findings reported in studies cited in the effects data base. Reburial and survival of amphipods was not significantly affected by sediments from the Palos Verdes site (Anderson et al. 1988). The total DDT concentration in the Palos Verdes sediment sample was 5,966 ppb. Marine worms that live in sediment appeared to be in excellent condition, and displayed normal burrowing behavior after 288 h of exposure to sediments containing 16,500 ppb DDT (McLeese et al. 1982).

A similar detailed analysis was done for the no-effects and effects total DDT data points used to derive the fresh water sediment TEL. The DDT analytical data for the first five data points at 0.384 ppb (Table 6) is not consistent with the major degradate being DDE. The major degradate of DDT in this study (Marking et al. 1981) was DDD. For example, the Red Wing Commercial Harbor sediments contained 5.28 ppb DDD, 0.56 ppb DDT and only 0.28 ppb DDE. The high proportion of DDD is contrary to the general finding that old residues of DDT are predominantly DDE. DDD is less stable in the environment than DDE.

The four no-effect data points at 5 ppb were all from nontoxic sediments, and the DDTs were not detected. The value of 5 ppb was chosen as one-half the detection limit of 10 ppb. Because the TEL is derived, in part, from the median of the no effects data set, these four data points with no detectable DDTs have much greater weight than do the last four data points containing thousands of parts per million DDTs, also with no effects.

The 200 ppb data point is one-half the detection limit of the unspiked control sediment used to determine the LC_{50} of DDT in the amphipod *Hyalella azteca*. The 1,300 and 1,800 ppb values are spiked sediments used to determine the LC_{50} of DDT in *Hyalella azteca* (Schuytema et al. 1989). These levels did not measurably affect the survival of this amphipod crustacean.

In the effects data set for total DDT in freshwater sediments, the study containing the 1.5 ppb value (Bolton et al. 1985) lists threshold contamination concentrations in Table 2.1 for sediments based on 4% organic carbon (OC) and an equilibrium existing between organic carbon and water. The threshold in water is the National chronic criterion. The OC is corrected to 1%. Only DDT is included. Total DDT from this study is not determined and is not used in the TEL data set. Notably, the threshold concentration for DDE (the predominant form of DDT in the environment) in this study is 28,000 ppb. The problem with these data is the apparent use of old K_{oc} values that give inaccurate estimates of the partition of DDTs between sediment organic carbon and water. The likely equilibrium partitioning threshold for DDT is higher than 1.5 ppb and lower for DDE and DDD at 7,000 ppb and 3,250 ppb at 1% OC, respectively.

The fresh water screening level concentration approach (SLCA) data point is 1.9 ppb. The salt water SLCA is 428 ppb and is based on sediments from the Southern California Bight. Neff et al. (1986) suggest that the difference is due to low DDT levels in the fresh water sediment data base and much higher DDT levels in the salt water sediment data base. Therefore, the difference appears to be an artifact of the method by which SLCA values are derived. The salt water SLCA was not used to derive the marine sediment TEL.

The fish tissue-based guidance of 5 ppb is derived from the equilibrium between water and sediment organic carbon, using a log K_{oc} of 5.92 (Hart et al. 1988).

The log K_{oc} of 5.92 is a geometric mean of values ranging from 5.26 to 6.58. The slow-stir log K_{ow} for DDT reported by de Bruijn et al. (1989) is 6.914. The K_{ow} derived by the superior slow-stir method gives a sediment acute marine threshold for DDT that is one order of magnitude higher.

The SLCA method that is used to derive the 7 ppb data point is the same as used by Neff et al. (1986) to derive the above mentioned value of 1.9 ppb. The method is described, but the actual data used to determine the 7 ppb value were not presented (Persaud et al. 1991).

The value of 9 ppb for DDT is for the parent compound and not total DDT. Actual data and calculations are not presented in the reference (Environment Canada 1992). The SLCA is not corrected for organic carbon. This data point, the 1.9 ppb data point (Neff et al. 1986) and the 7 ppb data point (Persaud et al. 1991) are all derived by the SLCA method, and are essentially the same, except for regional differences in sediment residue levels and biota. All three of these data points rely on mutual occurrence. None of them identify causality or represent a measure of dose-response to DDT.

The value of 50 ppb is reported to be the 90% effect level for the parent DDT according to the SLCA method (Environment Canada 1992). That is, 50 ppb of total DDT in sediments is associated with an effect on biota in 90% of those sediments. Any one or more of many hundreds of chemicals potentially present in those same toxic sediments could have accounted for the measured toxicity.

The 120 ppb effect data point is supposed to be the 95% effect level for the parent DDT according to the SLCA method (Persaud et al. 1991). This severe effect level is defined as that level "...that could potentially eliminate most of the benthic organisms." Any one or more of many hundreds of chemicals potentially present in those same toxic sediments could have accounted for the measured toxicity. The observation of apparently healthy benthic communities at sediment residue levels in excess of 120 ppb certainly puts in question the concept of this severe effect level for DDT as determined by the SLCA method.

According to the Illinois Environmental Protection Agency (IEPA 1988), the DDT effect level of 222 ppb is the average of three stations that had organisms of the lowest taxa. These three stations were also polluted by several additional contaminants, not just DDT. For example, Station GBL-08 sediments contained 270,000 ppb lead and 3,900 ppb mercury.

The value of 4,200 ppb is the calculated LC_{50} from the dose-response data from spiked sediment determination of the DDT LC_{50} for *Hyalella azteca* (Schuytema et al. 1989).

Discussion and conclusions. The TELs for total DDT are calculated from a variety of data types. Some sediment residue levels are considered to be background levels found in relatively unpolluted and nontoxic sediments; some are levels associated with toxic sediments; some are calculated from water column criteria and partition coefficients; some represent true dose-response from bioassays of spiked sediments. All of these data types should be considered in the determination of a sediment threshold for DDT toxicity. However, the TEL does not appropriately weigh the quality of the various data points. Outdated partition coefficients are included and should be removed or replaced with more accurate constants based on the slow-stir methodology (de Bruijn et al. 1989). Effects associated with relatively low concentrations of DDT are included, even though DDT concentrations that were orders of magnitude higher in sediments were without effect for the same biological endpoint. Bioassay data are given the same weight as all other data, even though bioassay data are the only data type representing true dose-response. Probably the most relevant data points of all, toxicity thresholds from bioassay data using spiked sediments, are under-weighted in the determination of TELs. Other troubling observations are the omission of data (even within the same studies), repeated use of the same data in different data points, the inconsistent correction for OC, and the use of data for just the parent compound in the determination of the TEL for total DDT. The only data points that address the issue of bioaccumulation beyond benthic organisms are the equilibrium-derived data points, but these appear to all have used older K_{oc} values that underestimate sediment thresholds.

Differences in freshwater and marine ecosystems can account for differences in toxic effect, but differences of several orders of magnitude are more likely explained by the presence of toxic levels of pollutants other than DDT in the sediments, rather than to inherent differences in the biological communities. How else can an effect level of DDTs at low parts per billion in one water body, and a no-effect-level of several thousand parts per billion in another water body be explained? The likely explanation is that other pollutants are present at toxic levels in the former and not in the latter.

The problem with using TELs as TMDL sediment targets is that risks are not accurately assessed, resulting in the assignment of resources disproportionate to risk and thereby not minimizing overall risk to humans and wildlife.

As an alternative to TELs, a sediment target for DDT can be derived that is based on the equilibrium between sediment and water as well as on bioaccumulation in fish and sensitive avian species. If one assumes a proportion of 80% DDE, 10% DDD and 10% DDT as an example of the residues typically found in sediments, the weighted log K_{oc} would be 6.77, using the slow stir K_{oc} values selected by the EPA in their 2002 DDT TMDL for Newport Bay and Watershed. Using this weighted K_{oc} value and the National criterion for the water column that is based on bioaccumulative avian toxicity, the resulting sediment target is 59 ppb.

3.2.2 NAS Fish Guidance to Protect Wildlife

The national debate over the impact of DDT on wildlife culminated in the U.S. cancellation of DDT in 1972. In the same year, the National Academy of Sciences (NAS) made recommendations for DDT residue levels in fish for the protection of wildlife (NAS 1972). One panel made a recommendation of 1,000 ppb in fresh water fish and another panel made a recommendation of 50 ppb in marine fish. The two panels cited many of the same scientific studies, in which eggshell thinning and reproductive failure were measured in sensitive avian species. Why then are the recommendations so different and which guidance is appropriate?

The recommendation of 1,000 ppb in fresh water fish appears to be based on laboratory studies in ducks and chickens, which are less sensitive species. The dose levels in these studies were intentionally high to assure inducing eggshell thinning and reproductive failure. The studies did not establish a chronic threshold for these effects. The panel admitted that their recommendation may not protect all species. The recommendation of 1,000 ppb in fresh water fish to protect wildlife appears not to be protective of reproduction in sensitive avian species.

The panel for marine fish guidance had one member, Robert Risebrough, who was an active investigator of the effects of DDT on eggshell thinning and reproduction in birds. At the time of this recommendation, Robert Risebrough had just published an article with Helen Hays on DDT in terns and fish scraps on Great Gull Island, 6 miles off the Connecticut coast in Long Island Sound. The 1970 fish data in this study (Table 7) appear to have become the basis for the 50 ppb recommendation. The data are reproduced below from the original publication by Hays and Risebrough (1972).

Table 7 DDT in fish from Great Gull Island. Table 2 in Hays and Risebrough (1972) reproduced with permission

HAYS AND RISEBROUGH
TABLE 2
DDT AND PCB RESIDUES IN FISH BROUGHT BY TERNS TO THE GREAT GULL ISLAND COLONY

Species	N	Mean weight (g)	ppm, fresh weight				ppm, lipid[1]		DDT/PCB
			p,p'-DDE	p,p'-DDD	p,p'-DDT	PCB	DDT	PCB	
Alosa aestivalis Blueback herring	5	12.4	0.22	0.18	0.011	0.64	6.4	10	0.64
Brevoortia tyrannus Atlantic menhaden	7	0.5	0.10	0.037	0.012	0.27	—	—	0.57
Clupea harengus Atlantic herring	2	3.3	0.022	0.027	0.00	0.38	—	—	0.13
Etrumeus teres Atlantic round herring	10	8.0	0.21	0.11	0.008	1.2	8.3	30	0.28
Anchoa mitchelli Bay anchovy	17	2.6	0.15	0.060	0.011	1.1	14	69	0.20
Menidia menidia Atlantic silverside	10	6.7	0.28	0.25	0.024	3.2	9.1	52	0.17
Morone americanus White perch	2	6.2	0.013	0.007	0.004	0.88	4.8	176	0.027
Scomber scombrus Atlantic mackerel	19	4.3	0.034	0.022	0.007	1.2	4.2	79	0.053

[1] Concentration of DDT is the sum of the concentrations of p,p'-DDE, p,p'-DDD and p,p'-DDT.

The recommendation of the panel is reproduced below from the NAS (1972) summary.

DDT Compounds

DDT compounds have become widespread and locally abundant pollutants in coastal and marine environments of North America. The most abundant of these is DDE [2,2-bis(p-chlorophenyl) dicholoroethylene], a derivative of the insecticidal DDT compound, p,p'-DDT. DDE is more stable than other DDT derivatives, and very little information exists on its degradation in ecosystems. All available data suggest that it is degraded slowly. No degradation pathway has so far been shown to exist in the sea, except deposition in sediments.

Experimental studies have shown that DDE induces shell thinning of eggs of birds of several families, including Mallard Ducks (*Anas platyrhynchos*) (Heath et al. 1969),[48] American Kestrels (*Falco sparverius*) (Wiemeyer and Porter 1970),[77] Japanese Quail (*Coturnix*) (Stickel and Rhodes 1970)[66] and Ring Doves (*Streptopelia risoria*) (Peakall 1970).[57]

Studies of eggshell thinning in wild populations have reported an inverse relationship between shell thickness and concentrations of DDE in the eggs of Herring Gulls (*Larus argentatus*) (Hickey and Anderson 1968),[36] Double-crested Cormorants (*Phalacrocorax auritus*) (Anderson et al. 1969),[31] Great Blue Herons (*Ardea herodias*) (Vermeer and Reynolds 1970),[70] White Pelicans (*Pelecanus erythrorhynchos*) (Anderson et al. 1969),[31] Brown Pelicans (*Pelecanus occidentalis*) (Blus et al. 1972;[36] Risebrough *in press* 1972),[62] and Peregrines (*Falco peregrinus*) (Cade et al. 1970).[37]

Because of its position in the food webs, the Peregrine accumulates higher residues than fish-eating birds in the same ecosystem (Risebrough et al. 1968).[64] It was the first North American species to show shell thinning (Hickey and Anderson 1968).[36] It is therefore considered to be the species most sensitive to environmental residues of DDE.

The most severe cases of shell thinning documented to date have occurred in the marine ecosystem of southern California (Risebrough et al. 1970)[64] where DDT residues in fish have been in the order of 1–10 mg/kg of the whole fish (Risebrough *in press* 1972).[62] In Connecticut and Long Island, shell thinning of eggs of the Osprey (*Pandion haliaetus*) is sufficiently severe to adversely affect reproductive success; over North America, shell thinning of Osprey eggs also shows a significant negative relationship with DDE (Spitzer and Risebrough, *unpublished results*).[78] DDT residues in collections of eight species of fish from this area in 1970 ranged from 0.1 to 0.5 mg/kg of the wet weight (Hays and Risebrough 1972).[47] Evidently this level of contamination is higher than one which would permit the successful reproduction of several of the fish-eating and raptorial birds.

Recommendation

It is recommended that DDT concentrations in any sample consisting of a homogenate of 25 or more fish of any species that is consumed by fish-eating birds and mammals, within the same size range as the fish consumed by any bird or mammal, be no greater than 50 μg/kg of the wet weight. DDT residues are defined as the sum of the concentrations of p,p'-DDT, p,p'-DDD, p,p'-DDE and their ortho-para isomers.

The DDT measured in the terns and in scraps of fish cast from their nests on Great Gull Island was not reported to have any effect on the terns. However, as detailed below, the breeding failure of ospreys along the Connecticut coast and on nearby Gardiner Island were clearly established in other studies. The implied assumption in the panel's recommendation is that the ospreys would be eating the same fish with the same level of residues found on Great Gull Island, and therefore that level was clearly toxic. What the summary didn't say was that the ospreys tend to feed along the coast and up the estuaries, resulting in a DDT exposure quite different from that of the terns. For example, osprey feeding patterns at a location further north were described by Greene et al. (1983). One could conclude from this and other studies that ospreys often catch fish from fresh or brackish water and, therefore, may not have been the best avian species for assessing the reproductive effect of DDT residues in marine fish. Fish from the nearby Connecticut River had much higher residues of DDT than the fish cast from tern nests on Great Gull Island.

Henderson et al. (1971) reported 0.85–3.27 ppm total DDT in fish collected from the Connecticut River in 1968 and 1969. In addition, Ames and Mersereau (1964) reported total DDT levels of 2.5–9.2 ppm in scraps of fish cast from osprey nests on Great Island near the mouth of the Connecticut River. Moreover, ospreys feeding in the Connecticut River estuary in 1967 were poisoned by dieldrin (Wiemeyer et al. 1975). These facts were known in 1972, but were not mentioned in the recommendation. The recommendation of 50 ppb does not appear to have taken into account all of the information available in 1972.

The above information sets the stage for considering the adoption of the NAS fresh water and marine fish recommendations for use today, more than 40 years later. Much has been learned about DDT and its effects on wildlife since 1972. The feeling among investigators in 1972 was concern, frustration, and even outrage at what was happening to avian species at the top of food chains. Within only a few years, however, recovery was well underway, and by 1980 was nearly complete in many species. The study of the recovery of the sensitive avian species indicates that toxicity thresholds do exist for DDT residues in fish diets. The results of such studies provide a way of observing dose-response over time as residues slowly declined. However, the relationship between fish and egg residues became less certain as levels in the United States declined below probable but unknown levels on wintering grounds in Latin America, where DDT use continued, even until today. The following chronology will focus on some of the key studies documenting the decline and recovery of the osprey, since this species is key to the NAS panel's recommendation for marine fish. Some of the earlier studies have been reviewed by Ware (1975).

Ames and Mersereau (1964) and Ames (1966) reported on the status of the osprey along the Atlantic coast. Most populations were experiencing dramatic declines associated with poor hatching and fledgling rates. Eggs and fish remnants from nests on Great Island at the mouth of the Connecticut River were assayed for DDT and metabolites in 1962. Eggs contained an average of 8.1 µg/ml (about 9 ppm fresh weight) total DDT and fish remnants cast from the osprey nests contained 2.5–9.2 ppm total DDT. A crude biomagnification factor would be $9/5.7=1.6$.

In 1963, Ames (1966) again studied osprey eggs from Great Island, and did a comparison with eggs from Maryland, where Ospreys were experiencing greater reproductive success. A few eggs from other locations along the Atlantic coast were also analyzed for DDT. The results are reproduced below in Table 8.

Table 8 DDT in osprey eggs from Northeastern United States. Table 2 in Ames (1966) reproduced with permission

Table 2. *DDT and its metabolites in the eggs of Ospreys from the north-eastern United States*

Locality	Year	No. of eggs	Average volume (ml)	DDE µg	DDE µg/ml	DDD µg	DDD µg/ml	DDT µg	DDT µg/ml	Total residues µg	Total residues µg/ml
Maine	1963	3	72	120	1·7	7	0·1	5	0·06	130	1·8
Rhode Island	1963	1	68	500	7·4	100	1·5	ND	ND	600	8·8
Connecticut	1962	6	68	450	6·7	100	1·5	Trace	Trace	550	8·1
Connecticut	1963	15	68	320	4·7	20	0·3	10	0·1	350	5·1
New Jersey	1963	2	Not measured	350	5·1	40	0·6	10	0·1	400	5·9
Maryland	1963	25	70	160	2·3	40	0·6	5	0·07	205	3·0

ND = None detected.

Table 9 DDT in fish from Connecticut and Maryland. Table 3 in Ames (1966) reproduced with permission

Table 3. *DDT residues in fish samples from Connecticut and Maryland*

Species	No. of individuals	Total wet weight (g)	DDE µg	DDE ppm	DDD µg	DDD ppm	DDT µg	DDT ppm	Total residues µg	Total residues ppm
CONNECTICUT										
Black-backed Flounder	6	376	160	0·4	30	0·1	300	0·8	490	1·3
Windowpane Flounder	2	70	50	0·7	10	0·1	140	2·0	200	2·9
Alewife	4	60	20	0·3	10	0·2	100	1·7	130	2·2
Shad	1	70	80	1·1	40	0·6	100	1·4	220	3·1
Cunner	1	19	Trace		Trace		60	3·1	60	3·1
Eel	1	40	80	2·0	40	1·0	100	2·5	220	5·5
MARYLAND										
Eel	4	572	60	0·1	110	0·2	60	0·1	230	0·3
Yellow Perch	3	256	20	0·1	10	0·04	30	0·1	60	0·2
White Perch	2	93	Trace		Trace		Trace		5	0·05
Striped Killifish	1	22	Trace		Trace		Trace		5	0·1
Menhaden	2	125	Trace		Trace		Trace		5	0·05
Toadfish	1	140	20	0·1	10	0·1	10	0·1	40	0·3

The Connecticut eggs contained an average of 5.1 µg/mL total DDT compared to 3.0 µg/mL in the Maryland eggs. Ames (1966) also collected fish from osprey nests in the Maryland and Connecticut studies (Table 9).

The Connecticut fish residues ranged from 1.3 to 5.5 ppm total DDT, whereas the Maryland fish residues ranged from 0.05 to 0.3 ppm total DDT. The differences shown in DDT levels in fish diet, and in eggs and reproductive success between the two colonies, is the first report of this kind. The results provide the first indications of the relationship between levels of DDT in the fish diet, in the egg, and hatching success. A crude biomagnification factor for Connecticut osprey in 1963, based on a weighted average fish residue of 2.1 ppm, is 5.7/2.1 = 2.7. For the Maryland data, again using a weighted average fish residue, a crude estimate of the biomagnification factor from fish to egg is 3.3/0.23 = 14. The increase in biomagnification factor with declining fish residues could result from the slow equilibration between dietary residues and adipose residues in the osprey and/or dietary sources higher in DDT than the fish that were measured. Because of the second possibility, greater weight should be given to fish data that were based on scraps from osprey nests. Even these data are subject to limitations, however, because what is measured is what the osprey didn't eat and often the remnants are dehydrated, resulting in higher residues than would exist for fresh weight. Other investigators also documented the decline in osprey populations.

Peterson (1969) reported on declining populations of ospreys in the United States and Europe. The declines were mostly the result of hatching failure and were attributed to pesticides. Henny and Ogden (1970) reported on the breeding success and status of osprey populations in seven states (Table 10).

Reese (1977) reported on productivity for osprey all across the United States for the period 1966–1974 (Table 11).

Table 10 Status of U.S. osprey populations. Table 1 in Henny and Ogden (1970) reproduced with permission

STATUS OF U.S. OSPREY POPULATIONS • *Henny and Ogden*

Table 1. The estimated present status of osprey populations in portions of seven states. The complete nesting populations of each state were not sampled, thus the total number of active nests presented in this table does not represent the size of the breeding populations and may not represent the status of the complete population in each state.

STATE	NO. ACTIVE NESTS (ALL YEARS SUMMED)	YEAR OF STUDY	NO. FLEDGED PER ACTIVE NEST	PERCENT NESTS SUCCESSFUL	ESTIMATED (MINIMAL) ANNUAL RATE DECLINE (PERCENT)	SOURCE OF NESTING STUDY
Florida	83	1968–69	1.22	70	stable	This paper
Minnesota	161	1966–68	1.03	65	2–3	Dunstan 1968
Maryland[b]	136	1964–65	1.03	54	2–3	Reese 1965
Wisconsin	128[a]	1952–59	0.98	53	3–4	Berger and Mueller 1969
Wisconsin	67	1960–65	0.39	30	12–13	Berger and Mueller 1969
Michigan	162	1965–67	0.39	23	12–13	Postupalsky 1969
Maine	8	1964	0.38	25[c]	12–13	Kury 1966
Connecticut	157	1960–63	0.29	23[d]	13–14	Ames and Mersereau 1964
Connecticut	30	1964–65	0.27	27[d]	13–14	Peterson 1969

[a] No data for 1957.
[b] Reese (Personal communication 1968) stated the first year of the study (1963) was preliminary and not as reliable as the following years. It was omitted.
[c] Kury (Personal communication 1969).
[d] Maximum percent of nests successful, assuming one young fledged per successful nest.

Table 11 Reproductive success in U.S. osprey populations. Table 8 in Reese (1977) reproduced with permission

JAN G. REESE
TABLE 8
RECENT NEST SUCCESS IN U.S. OSPREY POPULATIONS[1]

Location	Years	Nests	Nests Successful	Young Produced	Brood Size	Fledglings per nest	Reference
S. Massachusetts	1970–74	73	42	82	1.9	1.12	Fernandez (pers. comm.)
Chesapeake Bay:							
Eastern Bay	1966–74	323	128	229	1.8	0.71	Reese (1975)
This study	1970–74	684	386	741	1.9	1.08	
Choptank River	1968–74	188	106	190	1.8	1.01	Reese (1972, MS)
Smith Island	1968–71	71	55	98	1.8	1.38	Rhodes (1972)
Potomac River	1970–71	237	81	135	1.7	0.57	Wiemeyer (1971, 1977)
Virginia	1970–71	416	203	333	1.6	0.80	Kennedy (1971)
Michigan	1969–74	463	205	405	2.0	0.88	Postupalsky (1977 and pers. comm.)
Wisconsin	1966–69	237	111	193	1.7	0.81	Sindelar (1971)
Minnesota (Chippewa Nat. For.)	1968–72	249	120	216	1.8	0.87	Mathisen (1973)
Wyoming (Yellowstone Nat. Park)	1972–74	107	44	68	1.5	0.64	Swenson (1975)
Montana (Flathead Lake)	1967–70	80	42	77	1.8	0.96	Koplin (pers. comm.)
N. Idaho-E. Washington	1972–73	342	233	481	2.1	1.41	Melquist (1974)
Oregon (Deschutes Nat. For.)	1971	52	31	60	1.9	1.15	Lind (1971)
N. California	1969–71	136	71	139	2.0	1.02	Garber (1972)

[1] Data for all except this study were collected by two or infrequent nest visits and may not allow for mortality between final visit and fledging. Unpublished data are subject to revision.

Studies in other species soon identified eggshell thinning as the primary lesion causing hatching failure. DDE was shown to cause eggshell thinning in numerous declining species, including the osprey. Anderson and Hickey (1972) reported 21% shell thinning in osprey eggs collected in Connecticut, New Jersey and Maryland in 1957.

Johnson et al. (1975) reported 17% shell thinning in osprey eggs taken in Idaho in 1972 and 1973. Total DDT in eggs averaged 10.3 ppm. Hatching success was impaired. No fish residue measures were made. The general lack of use of DDT in the nesting grounds led the authors to suggest that exposure to DDT had occurred primarily during migration or at wintering grounds in Central America.

By 1973, fish residues, egg residues, eggshell thinning and hatching success appeared to be the critical determinants of the effect produced by DDE on osprey reproduction. All four parameters are highly correlated in declining species with exposures sufficient to cause eggshell thinning in excess of 10%. Mechanistic studies suggested that DDE acts directly on the transport, formation and/or deposition of calcium carbonate in the shell gland (e.g., see Risebrough et al. 1969).

Wiemeyer et al. (1975) evaluated known factors impacting reproduction in East Coast ospreys. The study period was 1968–1969. An egg exchange between nests in Maryland and Connecticut revealed that Connecticut eggs had lower hatching success than Maryland eggs, whether they remained in Connecticut or were moved to nests in Maryland. In contrast, Maryland eggs had higher hatching success than Connecticut eggs, whether they remained in Maryland or were moved to nests in Connecticut. The problem appeared to be the egg and not the parents or the setting. This finding is consistent with the direct effect that DDE has on the shell gland to produce thinner shelled eggs that were more susceptible to breakage, and therefore lower hatching success. DDE levels were higher in the fish diet of the osprey in some breeding areas than others, explaining the differential productivity along the East Coast.

Fish collected in Connecticut waters contained an average total DDT residue of 2.0 ppm. Fish collected in Maryland averaged 0.2 ppm total DDT. Fish scraps from osprey nests in Connecticut averaged 1.0 ppm, whereas one eel scrap from a nest in Maryland had 0.1 ppm total DDT. Fish scraps were judged to be very slightly dehydrated. Henderson et al. (1971) reported total DDT residues for 1969 in fish of 0.68 ppm for the Susquehanna River and 0.60 for the Potomac River. Both rivers flow into the Chesapeake Bay. Sampling locations on both rivers were in Maryland.

Total DDT in Connecticut osprey eggs collected in 1968–1969 was 10.3 ppm. This residue level compares with 10.9 ppm in 1964. Egg residues of total DDT from Maryland averaged 3.1 ppm. Eggshell thinning averaged 15% in Connecticut eggs and 12% in Maryland eggs. Only two eggs hatched out of 25 eggs studied in Connecticut. Fifteen eggs hatched out of 38 eggs studied in Maryland. Dieldrin may have contributed to hatching failure in Connecticut. Lethal concentrations of dieldrin were measured in a dead adult osprey found near the Connecticut River in 1967. Crude estimates of biomagnification from fish to egg were $10.9/(2.0 \text{ or } 1.0) = 5.4–10.9$ for Connecticut and $3.1/(0.68–0.1) = 4.6–31$ for Maryland.

In a 1972 study done on an offshore island along the Gulf coast of Florida, Szaro (1978) reported an average of 0.11 ppm total DDT in fish (lipid basis converted to fresh weight assuming 5% lipid), an average of only 0.43 ppm total DDT in eggs, a 9%

thinning of eggshells and 0.73 young per female. The lower than normal reproductive success was not attributed to the eggshell thinning, which was described as near normal. The fish were scraps taken from the same nests as the eggs. The fish muscle was analyzed. A crude biomagnification factor can be calculated as $0.43/0.11 = 3.8$. Whole fish would undoubtedly give a lower biomagnification factor. The population of ospreys in Florida is not migratory, remaining in Florida year-round.

Wiemeyer et al. (1978) reported on studies on osprey reproduction in New Jersey in the years 1970–1974. Until 1974, these breeding populations had high egg residue levels and poor productivity (Table 12). Fish residue data were not reported.

Eggshell thinning is summarized in Table 13.

Table 12 Average DDE residues in eggs and population status of osprey in New Jersey, 1970–1974. Data are from Table 2 in Wiemeyer et al. (1978)

Population	Year	p,p'-DDE (ppm wet weight)	Population trend and reproductive success
Potomic River, Maryland	1968–1969	2.4	Stable population; reproduction slightly depressed
Lake Coeur d'Alene, Idaho	1972–1973	8.5	Stable or increasing population; reproduction normal
Connecticut	1968–1969	8.9	Declining population; reproduction greatly depressed
Barnegat Bay Area, New Jersey	1974	16.0	Declining population; reproduction greatly depressed
Avalon-Stone Harbor, New Jersey	1970+1972	14.0	Declining population; reproduction greatly depressed

Table 13 Changes in shell thickness of New Jersey osprey eggs. Table 3 in Wiemeyer et al. (1978) reproduced with permission

TABLE 3
Changes in shell thickness of New Jersey osprey eggs.

Area	Year	Sample Size[a]	Average Shell Thickness $\pm$ 95% CL[b]	% Change from pre-1947
Eastern U. S.[c]	pre-1947	365 (-)	0.505 ± 0.004	--
Barnegat Bay Area	1971	2 (2)	0.485 ± 0.064 (0.48 – 0.49)	-4
Barnegat Bay Area[d]	1974	7 (4)	0.408 ± 0.073 (0.34 – 0.44)	-19
Avalon-Stone Harbor Area[d]	1970 + 72	8 (8)	0.443 ± 0.024 (0.40 – 0.49)	-12

[a] Number of eggs measured; number of clutches represented in parentheses.
[b] Means for current samples are on a clutch basis, while that for pre-1947 is on an egg basis. Complete clutches are usually represented in museum collections (pre-1947), whereas most recent samples are single eggs from clutches. Extremes of clutch means in parentheses. CL = confidence limits.
[c] From ANDERSON and HICKEY (1972).
[d] The eggs represented here are different in part from those that were analyzed for pollutants, as reported in Table 1; see text.

The first report of a significant recovery of ospreys was by Spitzer et al. in 1978, 6 years after the ban of DDT. Eggs collected from osprey populations in Connecticut and eastern Long Island from 1967 to 1970 had 15–20% thinning, approximating the critical level associated with hatching failure in other species. DDE levels in osprey eggs from this area declined fivefold between 1969 and 1976 and threefold between 1973 and 1976. "The productivity of these ospreys has since increased from about 0.5 fledged young per pair in 1969–1973 to 1.2 fledged young in 1976–1977 (Fig. 1), approaching the range observed in 1938–1942." The results are reproduced in Fig. 7.

Productivity improved when DDE residues in eggs fell below 12 ppm (60 ppm dry weight), a finding that is consistent with those of Henny et al. (1977) for other areas. The authors acknowledged that dieldrin probably affected survival and reproduction of ospreys in the Connecticut River estuary. No fish residue data were reported.

MacCarter and MacCarter (1979) reported improving reproduction in osprey at Flathead Lake in Montana even with high egg residues of DDTs, reproduced in Table 14.

From 1967 to 1977, the number of breeding adults gradually increased even though productivity was marginal as might be expected with the high levels of DDTs. Eggshell thinning and fish residues were not reported. The productivity data are reproduced in Table 15.

A report by Spitzer et al. in 1983 gave further indication of the recovery of osprey breeding along the northeastern coast as shown in Fig. 8.

The authors noted that DDE in osprey eggs had not been measured since 1976. Presumably DDE residues were declining as reproduction improved. They also made note of a brood-size reduction of 50% or more due to food limitations on Gardiner Island, the same island mentioned as impacted by DDT in the NAS recommendation for marine fish.

Spitzer and Poole (1980) and Poole (1989) revisited the issue of the struggling population of ospreys on Gardiner Island. The population was decimated by DDT in the 1950s and 1960s. Local citizens took up the cause to save the osprey. They sued Suffolk County to stop spraying DDT for mosquito control and achieved a ban on eastern Long Island. This local citizens group later became the Environmental Defense Fund. One of their members, Dennis Puleston, was an author of the 1978 report (Spitzer et al.) on the recovery of osprey populations on eastern Long Island. Recovery of the osprey on Gardiner Island was well underway in the 1970s when reproduction failed again due to a limited food supply. Apparently male osprey had to travel long distances to reliable supplies of fish in the marshes of the south fork of Long Island. According to the authors, when this colony thrived it was dependent on menhaden in nearby Gardiner's Bay. Excessive commercial fishing removed this food source, leading to a marginal food supply.

Reporting on a national survey of osprey breeding in 1983, Henny stated: "Ospreys at locations with poor production have all showed improvement following the DDT ban in 1972."

Wiemeyer et al. (1988) reported DDT effects on osprey eggs and reproduction from several data sets generated in the 1960s and 1970s. Some declines in residue

Fig. 7 Beginning of recovery of osprey productivity in Connecticut–Long Island. Figure 1 in Spitzer et al. (1978) reproduced with permission

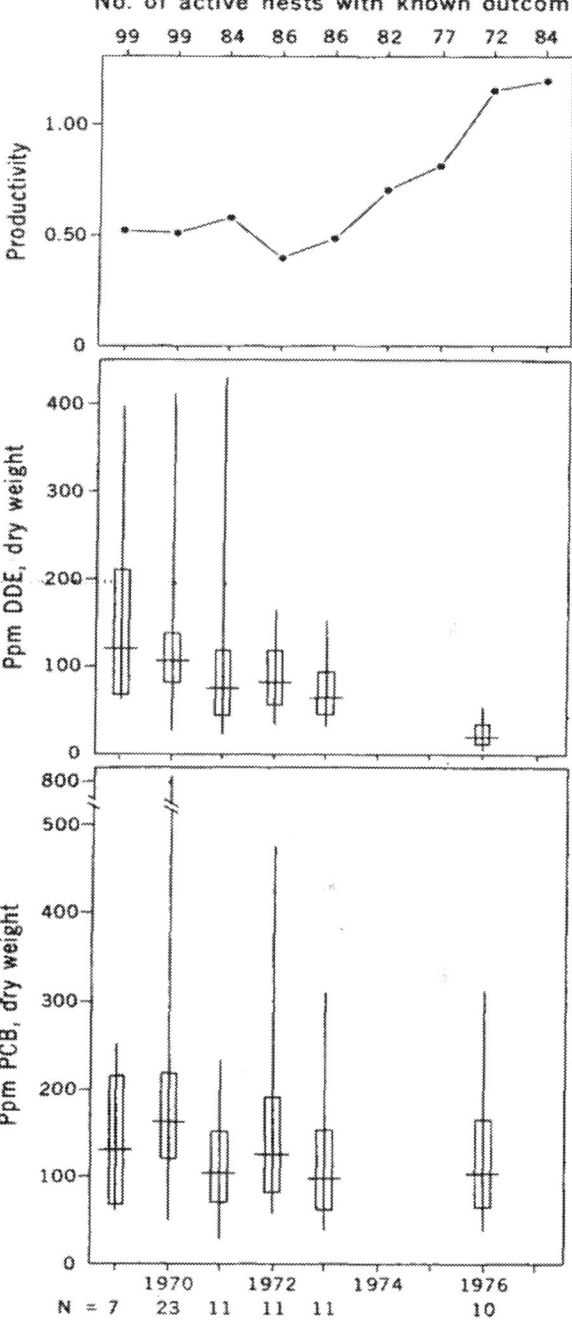

Fig. 1. Active nests of ospreys in Connecticut–Long Island with known outcome, 1969 to 1977; productivity, defined as young fledged per active nest; DDE and PCB residues, parts per million dry weight, with the sample sizes. Horizontal bars are geometric means; rectangles are the 95 percent confidence intervals of the means; vertical lines are the sample ranges.

Table 14 DDT residues in addled osprey eggs from Flathead Lake, Montana. Data are from Table 2 in MacCarter and MacCarter (1979)

Year	Nest number	Egg number	DDE[a]	DDD	DDT	Total DDTs
1968	BI-1	1	5.1	1.2		6.3
	BI-3	1	7.9	1.3		9.2
	DB-1	1	11.4	0.85		12.2
	DB-1	2	10.4	4.4		14.8
1969	BI-1	1	13.5	2.6		16.4
	BI-2	1	6.5	2.0		8.5
	BI-2	2	10.1			10.1
	DB-1	1	5.2			5.2
	DB-1	2	9.5			9.5
	BI-5	1	22.6			22.6
1970	BI-1	1	16.0		1.5	17.4
	BI-2	1	13.5		2.2	15.7
	BI-5	1	5.3		0.4	5.7
	BI-5	2	3.8			3.8
	N-D-1	1	5.9	0.6	1.7	8.2
1976	BI-3	1	3.1	0.12		3.22
	BI-5	1	37.0	3.3	0.35	40.65
	CB-8	1	35.0	5.6		40.60
1977	BI-3	1	2.9	0.14		3.04
	BI-5	1	16.0	1.2		17.20
	CB-8	1	8.7	1.1		9.8
	CB-8	2	11.0	1.2	0.20	12.40

[a]All residues are ppm wet weight

Table 15 Nesting productivity in osprey eggs from Flathead Lake, Montana. Table 3 in MacCarter and MacCarter (1979) reproduced with permission

TABLE 3. Nesting productivity of Ospreys at Flathead Lake, Montana.

Year	No. nesting pairs (A)	No. young (B)	No. young fledged (C)	No. nestlings per pair (B/A)	No. fledglings per pair (C/A)
1967	16	18	17	1.12	1.06
1968	20	14	14	0.70	0.70
1969	20	20	15	1.00	0.75
1970	24	33	31	1.38	1.29
1974	28	36	34	1.31	1.21
1975	30	41	38	1.37	1.27
1976	36	43	40	1.19	1.11
1977	38	38	36	1.00	0.95
Total Average	212	243	225	1.15	1.07

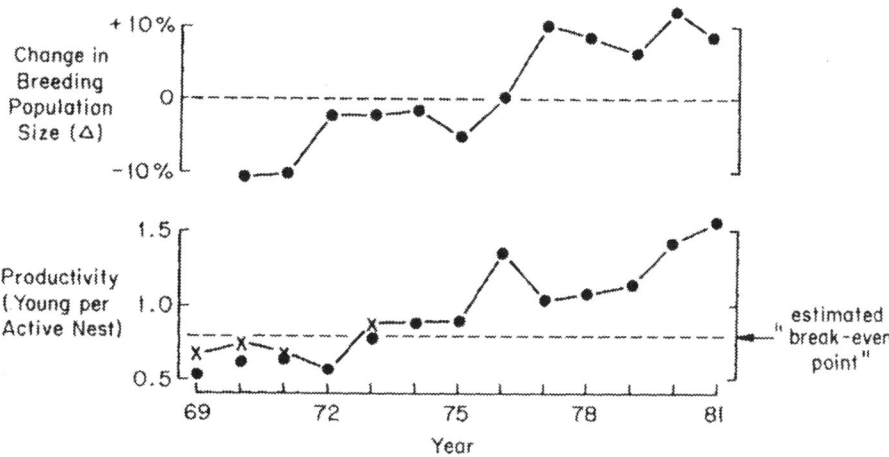

FIGURE 2. A comparison of Osprey reproductive rate and change in population size, N.Y. City to Boston, 1969-1981. Points denoted by "X" on the lower graph are productivity values which include young introduced from Maryland by Spitzer (1978).

Fig. 8 Productivity and population size of osprey in the region between New York City and Boston. Figure 2 in Spitzer et al. (1983) reproduced with permission

levels and shell thinning were noted. Analysis of the DDE egg residue—shell thinning relationship revealed 10% thinning at 2.0 ppm, 15% at 4.2 ppm and 20% at 8.7 ppm. Reproductive failure was attributed to DDE causing thinning of eggshells. Ospreys were considered to be as sensitive as other sensitive species.

In his book on ospreys, Poole (1989) published Figure 9.7 relating DDE residues in osprey eggs with eggshell thinning as shown in Fig. 9.

Poole's data illustrate the wide variability in eggshell thinning at each residue level, explaining why populations increase even at levels of DDE that result in some shells breaking and failing to produce viable young. Reproductive failure and mortality due to high residues of dieldrin and PCBs, particularly in the 1960s and early 1970s, may account for some of this variability. Poole also reported on the DDE egg residue—production dose-response as shown in Fig. 10. Poole set the reproductive effect threshold at 4.3 ppm DDE. This number compares with the 15% shell-thinning value suggested by Wiemeyer et al. (1988) at 4.2 ppm DDE.

Schmitt et al. (1990) published the results of a national fresh water fish residue survey for 1984. Total DDT residues in fish from the Connecticut River averaged 0.22 ppm. For all sites sampled nationwide, the trend of the geometric average total DDT residue was 0.39 ppm in 1976–1977, 0.36 ppm in 1978–1979, 0.32 ppm in 1980–1981 and 0.28 ppm in 1984. Schmitt et al. (1981) had earlier published a nationwide level of 1.08 ppm in fish collected between 1970 and 1974. Bilger et al. (1999) discussed EPA analysis of multi-species composite analyses done in 1987. The mean DDE concentration was 0.295 ppm in a nationwide sampling. The USGS

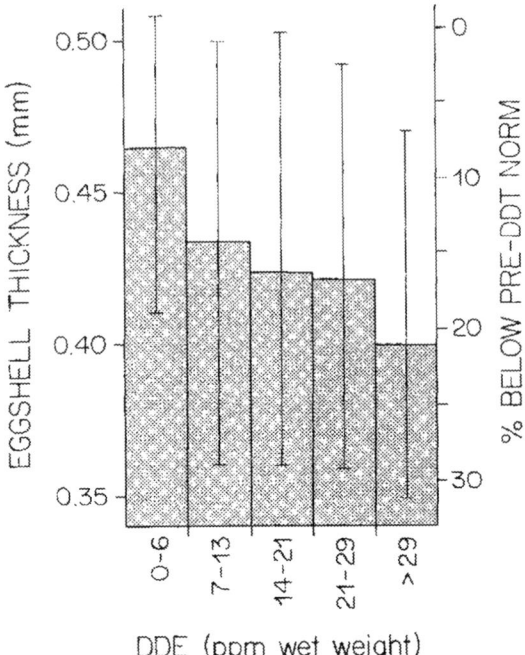

Fig. 9 Osprey eggshell thickness in relation to egg residue of DDE. Figure 9.7 in Poole (1989) reproduced with the permission from the author and from Cambridge University Press

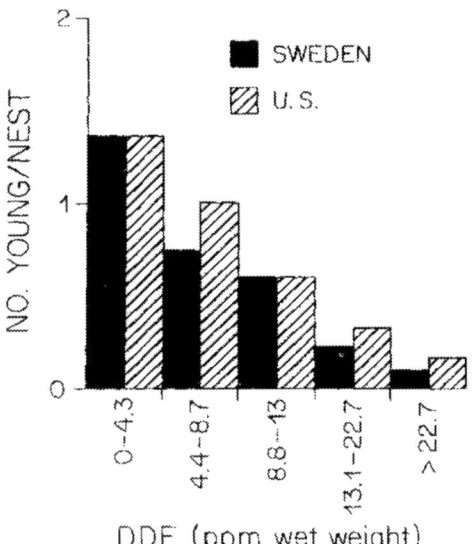

Fig. 10 Osprey productivity in relation to egg residue of DDE. Figure 9.8 in Poole (1989) reproduced with the permission from the author and from Cambridge University Press

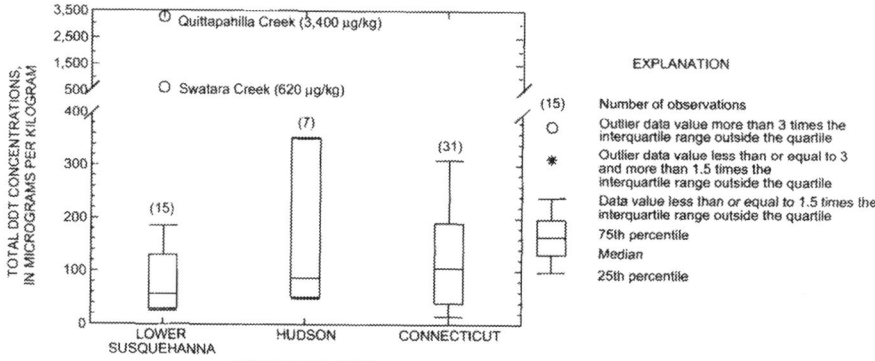

Figure 2. Concentrations of total DDT in white sucker whole fish tissue for the Lower Susquehanna, Hudson, and Connecticut River Basins.

Fig. 11 Total DDT in white sucker tissue from Northeastern United States. Figure 2 in Bilger et al. (1999) courtesy of the U.S. Geological Survey

multi-species fish sampling of the lower Susquehanna River basin by Bilger et al. (1999) in 1992 found median residues of 0.250 ppm of total DDT. Variability between sites was very high as shown in results for white suckers collected from the Susquehanna, Hudson and Connecticut River Basins (Fig. 11).

The overall trend for DDT in fish residues in the 1970s and 1980s was a steady decline, although hot spots were clearly evident. If these hotspots are sources of food for ospreys and are missed in fish surveys, then the residue exposures may be greatly underestimated, resulting in an overestimate of biomagnification from fish to osprey egg.

Steidl et al. (1991a, b) published two papers on osprey reproduction in three regions of southern New Jersey. The three locations were the more polluted Delaware Bay, the less polluted Atlantic Coast and an intermediate location along the Maurice River that flows into the lower Delaware Bay. Eggs were collected in 1985–1989. The authors noted that average fish residue of total DDT in the Delaware River was 0.88 ppm in 1984. Total DDT residues in eggs were low with the highest levels in Delaware Bay as shown in Table 16. Eggshell thickness was negatively correlated to DDE levels in the eggs as can be judged from Tables 16 and 17.

Apparently, eggs with shells thinned near to or at 15% had a greater probability of breaking, contributing to the lower productivity observed in Delaware Bay compared to the other two locations as shown in Table 18.

Analysis of DDT residues in prey fish revealed the following results as shown in Table 19.

One should keep in mind that these fish samples were not scraps from the osprey nests but fish caught locally in the breeding grounds. Since ospreys often feed up the rivers from their breeding grounds, more contaminated fish may well have been consumed. Viscera were removed from whole fish prior to analysis. Viscera would contain liver, some adipose tissue and other organs that would be expected to have

Table 16 Geometric mean DDE and DDD residues (ppm wet wt) in osprey eggs from Delaware Bay. Data from Table 1 in Steidl et al. (1991b)

Region and egg type	n	DDE (geometric mean)	(range)	DDD (geometric mean)	(range)
Delaware Bay					
Random	7	3.2	1.7–5.2	0.4	0.3–0.7
Addled	4	2.9	1.6–4.7	0.4	0.3–0.6
All	11	3.1		0.4	
Atlantic Coast					
Random	8	1.2	0.5–2.8	0.2	0.1–0.6
Addled	4	1.6	1.4–1.8	0.2	0.2–0.3
All	12	1.4		0.2	
Maurice River					
All	2	1.9	1.6–2.3	0.2	0.2–0.2

Table 17 Eggshell thickness of random (1989) and addled (1985–1988) osprey eggs, and eggshell fragments (1987–1988), from three regions of New Jersey. Data from Table 3 in Steidl et al. (1991b)

Region and shell type	n	Eggshell thickness (mm) Mean	SE	% Below pre-1947 thickness[a] Mean	SE
Delaware Bay					
Random	7	0.444	0.020	12.0	3.9
Addled	8	0.466	0.014	7.8	2.7
Fragment	2	0.430	0.005	14.9	1.0
All types	17	0.453	0.011	10.4	2.1
Atlantic Coast					
Random	8	0.485	0.020	4.0	3.9
Addled	22	0.488	0.011	3.3	2.2
Fragment	19	0.472	0.011	6.5	2.1
All types	49	0.482	0.007	4.7	1.4
Maurice River					
Random	2	0.490	0.045	3.0	8.9
Fragment	2	0.465	0.005	7.9	1.0
All types	4	0.478	0.020	5.5	3.9

[a]Compared to data from Anderson and Hickey (1972)

relatively high concentrations of DDT. Finally, these fish were caught in 1989 and the eggs were collected from 1985 to 1989. Some decline in fish residues from 1985 to 1989 would be expected, based on data from other locations. Given all of the above, crude bioconcentration factors can be calculated as 5.7/0.54 = 5.7 for the Delaware Bay, 1.4/0.09 = 15.6 for the Atlantic coast and 1.9/0.095 = 20 for the Maurice River.

As the level of DDT in fish decreased, the bioconcentration factor appears to have increased. This pattern became even more evident as fish residues continued to

Table 18 Osprey productivity in three regions of New Jersey. Table 2 in Steidl et al. (1991b) reproduced with permission

Table 2. Reproductive parameters of ospreys nesting in 3 regions of New Jersey, 1987–88.

Region	n	% eggs hatched	$\bar{x}$ young fledged/pair	% nest success[a]
Delaware Bay[b]	24	50.0[c]	1.08	50.0
Atlantic Coast[b]	38	68.5	1.61	78.9
Maurice River	6	62.5	1.33	66.7

[a] Nests fledging ≥1 young.
[b] Data from Steidl et al. (1991).
[c] n = 12 nests.

Table 19 DDE and DDD (ppm fresh wt) in osprey prey fish collected from three regions of New Jersey. Data from Table 5 in Steidl et al. (1991b)

Region and species	n[a]	DDE	DDD	% Moisture
Atlantic Coast				
Menhaden	5	0.05	0.04	62.8
Delaware Bay				
Menhaden[b]	5	0.17	0.12	66.2
White perch[c]	5	0.68	0.27	71.8
Channel catfish[d]	2	0.25	0.14	71.4
Maurice River				
White perch	6	0.05	0.03	72.2
Channel catfish	2	0.08	0.03	76.0

[a] Number of fish in composite sample
[b] Composite contained (ppm) 0.02 p,p'-DDT, 0.03 o,p'-DDE, 0.08 o,p'-DDD
[c] Composite contained (ppm) 0.11 o,p'-DDE, 0.27 o,p'-DDD
[d] Composite contained (ppm) 0.03 o,p'-DDE, 0.05 o,p'-DDD

decrease. One must keep in mind that as DDT residues continued to decrease in the United States, following the ban in 1972, exposure to DDT in wintering grounds in Latin America accounted for an increasing proportion of egg residues. DDT was used in Latin America after 1972 and is still in use in some locations today. These wintering ground exposures became more important as residues in fish in the U. S. continued to decrease. The multi-year half-life of DDT ensures that the highest exposures will be reflected in adipose concentrations that are passed directly into the yolk of the egg.

Other factors contributing to reproductive effects in ospreys in southern New Jersey include the presence of 4.1–26 ppm PCBs in the osprey eggs from Delaware Bay (Steidl et al. 1991b). The authors noted that the Delaware Bay is routinely dredged to maintain a shipping channel to ports on the Delaware River. They suggested that dredging exposed biota to old sediments containing higher residues of DDT and PCBs, resulting in a slower decline of residues and the persistence of effects no longer seen at other locations. Another factor is the travel time required to catch fish due to the lack of clarity of the water in the nesting areas that are in the more polluted parts of the Bay. Long travel times did not limit the food supply but did increase the

Table 20 The importance of wintering ground exposures to DDT in peregrine falcons in the late 1970s. Table 1 in Henny et al. (1982a) reproduced with permission

TABLE 1. DDE (geometric means, ppm wet weight) in blood plasma of Peregrine Falcons captured during migration at Assateague Island, Maryland/Virginia and Padre Island, Texas.

Year	Maryland/Virginia[a]			Texas			Maryland/Virginia			Texas		
	Mean	(95% C.I.)	n	Mean	(95% C.I.)	n	Mean	(95% C.I.)	n	Mean	(95% C.I.)	n
	HY♀♀						HY♂♂					
1976-77	0.11	(0.07–0.19)	15	0.05	(0.03–0.08)	15	0.08	(0.05–0.14)	9	0.16	—	2
1978	0.04	(0.02–0.08)	25	0.03	(0.02–0.07)	20	0.03	(0.01–0.10)	8	0.06	(0.03–0.09)	16
1979	0.07	(0.05–0.10)	36	0.05	(0.04–0.07)	74	0.08	(0.06–0.11)	26	0.05	(0.04–0.08)	22
Totals	0.06	(0.05–0.09)	76	0.05	(0.04–0.06)	109	0.07	(0.05–0.09)	43	0.06	(0.04–0.08)	40
	SY♀♀						ASY♀♀					
Fall												
1976-78	0.82	(0.44–1.53)	11	0.28	(0.01–6.75)	4	—			0.60	(0.27–1.33)	6
1979	0.64	(0.38–1.07)	6	0.27	(0.02–3.91)	3	0.71	(0.14–3.67)	5[b]	0.33	(0.14–0.77)	12
Totals	0.75	(0.50–1.13)	17	0.28	(0.07–1.16)	7	0.71	(0.14–3.67)	5	0.40	(0.22–0.72)	18
Spring												
1978-79	—			1.43	(0.52–3.87)	8	—			0.88	(0.60–1.29)	21
1980	—			0.42	(0.24–0.73)	19	—			0.62	(0.48–0.79)	63
Totals	—			0.60	(0.36–1.00)	27	—			0.67	(0.55–0.83)	84

[a] Excludes 3 HY♀♀ that were released along East Coast by Cornell University biologists.
[b] Includes one sample from 1978.

time the nests were unattended, leading to potentially greater predation by great horned owls.

Considering the importance of the unknown exposure of ospreys to DDT in wintering grounds, digression to an article by Henny et al. (1982a) is enlightening. Henny et al. reported the measurement of DDT in the blood of peregrine falcons captured during migration north in the spring and south in the fall (Table 20). The peregrine falcon migration is similar to that of the osprey.

The table requires explanation. HY falcons are those migrating in the year they hatched. SY falcons are second year falcons and ASY means falcons migrating after their second year. Focusing on the Texas data for female falcons, one can see that just fledged falcons on their way south have quite low levels of DDE. SY falcons returning north in the spring of the next year have more than ten times as much DDE in their plasma. Plasma levels are lower in SY falcons migrating south from northern breeding areas. Apparently, body burdens gained in the south during the winter are decreasing in the north during summer as a result of both egg laying and ever decreasing exposures in the northern breeding areas. The same pattern would be expected in osprey.

This exposure paradigm is even more important for the osprey since fledglings do not return to northern breeding grounds until their third year. Southern exposures to DDT could explain the ever increasing bioconcentration factors calculated from just northern exposures. As DDT levels decreased in the United States to levels below those in Latin America, the importance of the unknown southern exposure eventually becomes essential to understanding the relationship between DDE residues in the fish diet and levels in eggs associated with thinning and hatching failure. With the understanding gained from this digression, let us resume reviewing the chronology of studies of the effects of DDE on osprey reproduction.

Audet et al. (1992) measured DDT residues in osprey eggs from three locations on the East Coast and compared them with residue levels in the early 1970s. The study was prompted by the finding of an isolated area in Chesapeake Bay with declining nestling survival. Median DDE levels in 1986 were 2.3 ppm in an area of declining fledgling survival (Martin Refuge), 0.65 ppm in coastal Virginia and 0.56 ppm in southern coastal Massachusetts. Relatively high ratios of DDT to DDE in the eggs from Massachusetts prompted the authors to suggest recent exposure to DDT from an unknown source (winter breeding grounds?). Eggs taken in 1972–1973 from the same area of Massachusetts had DDE residues of 4.2 ppm. The authors concluded that the 0.65 and 0.56 ppm levels of DDE "were well below reported values associated with biologically significant effects on eggshell thickness and reproductive success." In 1973, the Martin refuge had a median DDE level in eggs of 3.4 ppm with 17% eggshell thinning, but nonetheless, 1.5 young per active nest. Productivity of 1.5 young per active nest was considered by these authors to be excellent. No reason was given or suggested for the declining fledgling survival at the Martin Refuge in 1986. Fledgling survival data were not reported.

Falkenberg et al. (1994) provided data on a nonmigratory population of osprey and their prey from the south coast of Australia. Six eggs collected in 1987 had an average total DDT residue of 0.22 ppm. Total DDT residues in three species of prey fish averaged 0.3 ppm, giving a very low biomagnification factor of 0.73. Shells of osprey eggs collected in 1987–1988 were no thinner than shells of eggs collected prior to the DDT era. The biomagnification of DDT into osprey eggs was so low in this study as to put into question the representativeness of the fish samples as a significant part of the diet eaten by osprey that produced the eggs collected in the study. The determination of a biomagnification factor is theoretically more certain in a nonmigrating population. Most likely, the biomagnification factor is small, based on studies in the 1960s and early 1970s in the U. S., probably less than ten.

In 1997, Ewins published an article about the behavior and history of osprey in North America. Figure 1 from Ewins (1997) illustrates the recovery of ospreys in Wisconsin and the Georgian Bay area of the Great Lakes Region (Fig. 12).

Woodford et al. (1998) reported geometric mean DDE residues of 0.20–0.52 ppm in osprey eggs collected in 1992–1993 from two breeding areas in central and northern Wisconsin.

Ewins et al. (1999) reported on eggs collected between 1980 and 1989 from two osprey breeding areas in central Michigan. The known age of each female osprey producing the eggs permitted a study of DDT residues in eggs produced by females from 3 to 15 years of age. No age-related changes were found. The egg residues were independent of the age of the female, and DDE averaged 1.2 ppm. Eggshell thickness increased from 1980 to 1989. Eggs collected from 1980 to 1984 were 5% thinner and eggs collected from 1985 to 1989 were 3% thinner than eggs collected prior to the DDT era. Eggs collected from the same areas in 1972–1973 had geometric mean concentrations of 5.1 ppm DDE and 10% average shell thinning. The decrease in DDE residues was associated with improved reproduction and population increases. Apparently, female osprey in Michigan reached a steady-state DDE residue level in their tissues in the first 2–3 years of life (most of that time is spent

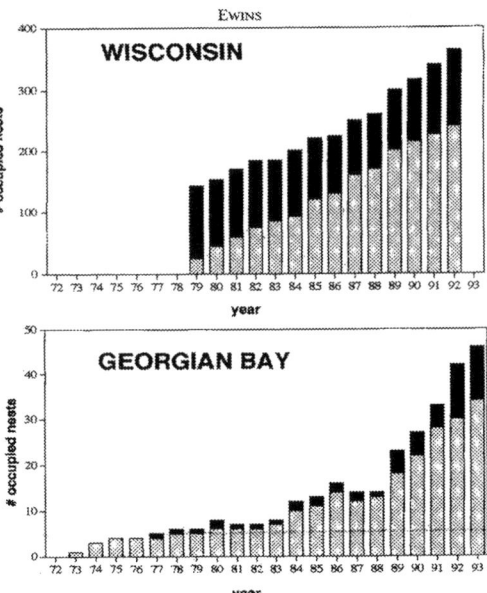

Figure 1. Changes in breeding populations of Ospreys since early 1970s in Wisconsin and Georgian Bay (Lakes Huron and Ontario), at artificial nest-platforms (stippled) and other (solid shading) sites. Most "other" sites were in trees. Wisconsin data are from Gieck et al. (1992).

Fig. 12 Recovery of ospreys in Wisconsin and the Georgian Bay area of the Great Lakes. Figure 1 in Ewins (1997) reproduced with permission

on the wintering grounds in Latin America). Part of this ongoing steady-state is the elimination of accumulating adipose residues by laying eggs. The DDE residues in eggs from midwestern breeding grounds and some east coast locations appear by the mid 1990s to be below levels associated with any significant effects on shell thickness or hatching success.

Elliott et al. (2000) reported on DDT residues in osprey eggs collected from the Columbia and Fraser River areas in the northwest. DDE residues were high and variable. Geometric means ranged from 1.0 to 13.8 ppm by area and year from 1991 to 1997. No trends by area or year were evident. Individual eggs ranged from 0.1 to 23.7 ppm DDE. DDE/DDT ratios were also highly variable. Some of the locations were in forested wilderness areas where little DDT had been used. Fish sampled in 1994 from these remote areas contained less than 0.005 ppm total DDT. The authors suggested that DDT was coming from an outside source, possibly from wintering grounds in southern Mexico. Another factor is the very high rate of DDT applications to apple orchards during the DDT use era (Blus et al. 1987). Twenty three percent of the osprey eggs had DDE residues greater than 4.2 ppm, the level associated with eggshell thinning that is significant to hatching success.

Clark et al. (2001) published a follow-up study of the Steidl et al. (1991b) Delaware Bay study summarized above. Comparisons between 1989 and 1998 at three locations in southern New Jersey were made in residue levels in eggs and fish, eggshell thinning, and productivity. DDE residues in osprey eggs had declined to 1.4 ppm with an associated eggshell thinning of 7% in the more contaminated Delaware Bay area. The authors concluded that "PCBs and DDE in osprey eggs were below levels considered to be toxic to egg development." Fish were collected in the same manner as in 1989. Total DDT residues in fish for the Delaware Bay averaged 0.23 ppm. Biomagnification factors from fish to eggs ranged from 9 to 11. Osprey productivity increased to 1.1 young per nest in the period from 1994 to 1998. Availability of nest structures and owl predation were thought to be limiting the population of ospreys in the Delaware Bay area.

In 2003, Martin et al. reported on ospreys in Great Lakes Canada. The study was conducted in 1991–1995. DDE levels averaged 1.3–2.9 ppm in five study areas. A few eggs exceeded the 4.2 ppm (15% eggshell thinning) threshold, suggesting that reproduction in a few individual ospreys was affected. The authors concluded, however, "…ospreys now appear to be relatively unaffected by current low levels of chlorinated hydrocarbon contaminants."

Henny et al. (2003) reported on a detailed 1993 study of bioaccumulation of DDE from fish to osprey eggs in Oregon. The number of breeding pairs along the Willamette River increased from 13, in 1976, to 78, in 1993 and 234 in 2001. Overall productivity was 1.67 young per active nest. The geometric mean of DDE egg residues was 2.3 ppm. Two of the ten eggs analyzed had levels of DDE that would be expected, based on other studies, to have reduced hatching success as a result of cracked shells.

The median level of DDE in the major food fish for ospreys, the largescale sucker, was found to be only 0.022 ppm. This very low fish residue resulted in a bioaccumulation factor for fish to osprey eggs of 87, prompting the authors to suggest that ospreys received significant exposures during winter migration to southern Mexico and Central America. This idea was reinforced by lower than expected bioaccumulation of PCBs and unexpectedly high levels of DDT in some eggs. However, much higher levels of DDE in largescale suckers from the Willamette River have been reported. A single composite collected between 1996 and 1998 contained 0.835 ppm (US EPA Region X 2006). A bioaccumulation factor from this fish residue value would be $2.3/0.835 = 2.8$.

In a chapter in Raptors Worldwide, Henny et al. (2004) described a study of the effects of DDE residues on osprey eggshells and reproduction at nest sites along the Columbia River in northwestern United States. The number of ospreys had been increasing with each survey through 1998. Mean productivity was 1.64 young per active nest (Table 21). Eggs were collected in 1997 and 1998.

Dividing the nests into three classes by DDE egg residue level indicates a dose-response for thinning of eggshells and impairment of reproduction. Even at these high levels, with measurable impacts, the osprey population continues to grow. The geometric mean residue of DDE in eggs from nests along the Columbia River was 4.9 ppm, a value higher than residues reported by the same authors for eggs collected in 1993 along the adjoining Willamette River. These residues are the highest reported nationwide for osprey eggs during the late 1980s and 1990s. Henny et al. suggested the possibility of exposure to DDT on the wintering grounds in

Table 21 Productivity, eggshell thinning and DDE egg residues in osprey along the Columbia River. Table 5 in Henny et al. (2004) reproduced with permission

Table 5. Number of young Ospreys produced per nest (with one egg collected) in relation to DDE concentrations in the sample egg collected, and eggshell thickness.

	Number of Nests with DDE ($\mu g\ kg^{-1}$)		
Number of Young	< 4200	4200-8000	> 8000
0	1	3	3
1	6	3	2
2	10	6	3
3	1	0	0
Active Nests	18	12	8
Successful Nests	17	9	5
Adv. Young	29	15	8
Young/Successful Nest	1.71	1.67	1.60
Young/Active Nest	1.61	1.25	1.00
Geo. Mean DDE ($\mu g\ kg^{-1}$)	2131	5473	10510
Mean Shell Thickness (mm)	0.488	0.441	0.419
Shell Thinning[a]	-3.4%	-12.7%	-17.0%

Note: One nest sampled did not have complete information for productivity (it was excluded), and 10 nests were included from the Willamette River in 1993 (Henny and Kaiser 1996).

[a] Compared to 0.505 mm for pre-DDT era eggshells from eastern U.S.A. (Anderson and Hickey 1972).

southern Mexico and Central America. Another explanation could be the high application rates of DDT to apple orchards, creating pockets of high residues in soil and biota, including fish (Blus et al. 1987).

Fish residues were stated to be elevated, although levels were not reported. Previous investigations from 1991 to 1993 were cited by the authors to have found an average of 0.089 ppm DDE in largescale suckers, an important food fish for the ospreys. Schmitt et al. (1990) had reported 1.0 ppm total DDT in largescale suckers from the Columbia River in 1984. The US EPA Region X (2006) reported average total DDT residues of 0.450 ppm in largescale suckers collected in 1996–1998 from the Columbia River Basin. Figure 2-4b from the report (Fig. 13) illustrates the high variability in the fish residues at different locations, explaining to some degree the high variability in DDE levels in osprey eggs. A crude estimate of the biomagnification of DDE from fish to egg would be $4.9/(0.450–0.089) = 11–55$.

Martell et al. (2001) used satellite telemetry to track the migration of osprey from northern breeding areas to southern wintering areas. East coast osprey winter primarily in Brazil, west coast osprey winter primarily in southern Mexico and midwestern osprey winter in both locations or in between (Fig. 14).

Mora (1997) reviewed available information on reports of DDT contamination of migratory birds in Mexico. Contamination generally was found to be similar to that in southwestern United States through the 1980s.

Rattner et al. (2004) reported on contaminant exposure and reproductive success of ospreys in the Chesapeake Bay area. From a population estimated at 1,450 nesting pairs in 1973, the Chesapeake Bay osprey population more than doubled to an estimated 3,473 pairs by 1995–1996. However, reproduction rates have not fully recovered in the more polluted waters of the Bay. Geometric means of DDE levels

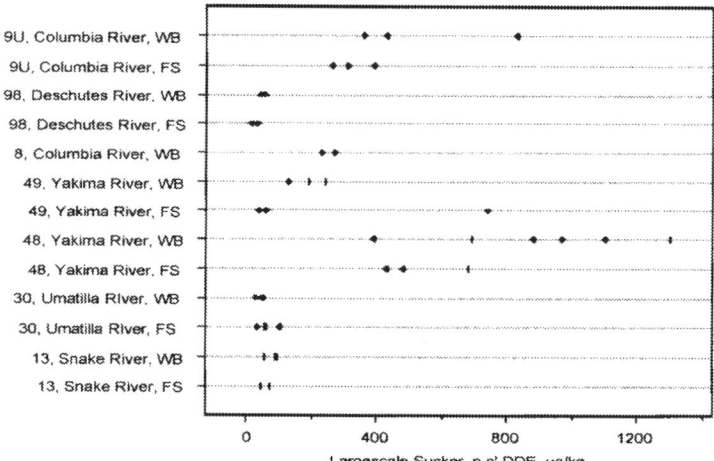

Figure 2-4b. Study site specific concentrations of p,p DDE in largescale sucker composite fish tissue samples from the Columbia River Basin.

Fig. 13 DDE residues in largescale suckers from the Columbia River Basin. Figure 2-4b reproduced from US EPA Region X (2006)

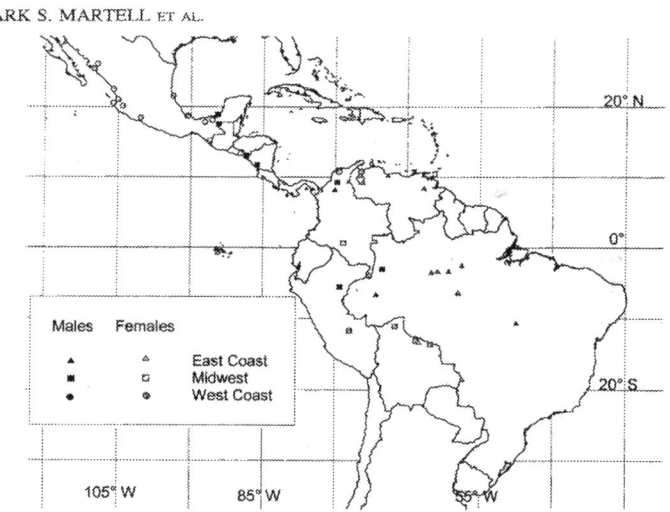

FIGURE 2. Wintering locations of North American Ospreys as determined by satellite telemetry.

Fig. 14 Wintering locations of North American Ospreys. Figure 2 in Martell et al. (2001) reproduced with permission from author and from The Condor, published by the Cooper Ornithological Society

in eggs collected in 2000 from different parts of the Bay ranged from 0.4 to 1.2 ppm. Eggshell thinning ranged from 0 to 9%. PCBs were as high as 19 ppm in eggs from nests in the more polluted areas. A limited sample of fish scraps from nests in some of the less polluted areas contained less than 0.05 ppm DDTs, corresponding to

0.4–0.8 ppm total DDT in osprey eggs from those areas. A crude estimate of the bioaccumulation of total DDT from fish to eggs would be 0.4–0.8/<0.05=>8 to >16. Marginal productivity in the more polluted areas was not linked to egg concentrations of DDE. DDE levels in osprey eggs from the Chesapeake Bay have decreased tenfold from the DDT use era. "…concentrations of p,p'-DDE…in sample eggs did not cause direct and biologically significant toxic effects on osprey reproduction in Chesapeake Bay regions of concern."

A third study of the Delaware Bay was conducted in 2002 by Toschik et al. (2005). Geometric mean DDE levels in eggs from four parts of the Bay were 0.4–1.8 ppm. Eggshells from the northern part of the Bay were 10% thinner. A few eggs from failed nests contained more than 4 ppm DDE. The authors stated that "All nestlings appeared in good health; no external lesions or other abnormalities were found." and "Additionally, no evidence of chromosomal damage in nestlings was found." Based on only a few eggs, DDE in eggs from the Prime Hook National Refuge were 0.6 ppm in 2002 compared to 5 ppm in 1974. Marginal reproduction rates in the more polluted areas were the result of lost eggs. Lost eggs can be the result of damaged or cracked eggs tossed out by the parents, eggs lost from precarious nests (e.g., on floating buoys), human interference, or predation. Some of these factors are more prevalent in the more polluted areas because they are also the more urbanized and industrialized areas. The authors concluded that "…the latitudinal trends seen in egg contaminant exposure are unlikely to result from contaminant exposure on the wintering grounds." This idea is somewhat contradicted by the wide range of DDE levels in eggs from each area (overall range of 0.17–4.61 ppm). No fish residues were reported.

The final osprey breeding study to be reviewed is one currently underway in Newport Bay. Newport Bay has historically seen osprey stopping briefly during migration. Ospreys prefer coniferous snags (e.g., dead fir trees) for nesting to minimize predation as well as to have a clear path for flying to and from the nest. The snag must also be close to fishing grounds so that a nesting pair has ample access to food for themselves and their young. Despite excellent fishing grounds, Newport Bay has historically not had this type of tree for nesting ospreys. In 1993, a wooden platform was affixed to the top of a 50 ft pole on Shellmaker Island in an attempt to attract an osprey pair to nest. Nothing happened for more than a decade. Sticks were added in an attempt to start the nest building activity. A pair built a nest in 2005. Mating resulted in two nonviable eggs. The next year, mating resulted in the successful production of two fledglings. Breeding has continued successfully each year by this pair. Tagging of the fledglings began in 2008. A female osprey, fledged at Shellmaker Island in 2008, returned to the area in 2010 with a mate and built a nest on a new platform at the San Joaquin Wildlife Sanctuary (SJWS) in the lower drainage into Newport Bay. This pair produced a single fledgling in 2010. In 2011 the same pair was in the process of raising two chicks, when suddenly both died of unknown cause. Hemorrhaging was noted in their lungs. The adults appeared unharmed and returned in 2012 to successfully fledge two chicks. The reproductive data for the two osprey pairs are listed in Table 22 below.

Table 22 Reproduction of osprey pairs at Shellmaker Island (SI) and San Joaquin Wildlife Sanctuary (SJWS)[a]

Year	Site	Eggs	Chicks	Fledglings	Productivity	Comments
2005	SI	2	0	0	0	First mating
2006	SI	≥3	3	2	2	Egg count not available
2007	SI	≥3	3	2	2	Egg count not available
2008	SI	≥3	3	3	3	Egg count not available
2009	SI	≥4	4	4	4	Egg count not available
2010	SI	3	3	3	3	
2010	SJWS	1	1	1	1	Female SI 2008; first mating
2011	SI	≥3	3	3	3	Egg count not available
2011	SJWS	≥2	2	0	0	Chicks died of unknown cause; hemorrhaging noted in lungs
2012	SI	≥3	3	3	3	Egg count not available
2012	SJWS	≥2	2	2	2	Egg count not available

[a]Data from Thomas (2010), Reed (2010), Reicher (2010), Kerr (2006) as well as personal communications from: Carla Navarro Woods, Reserve Manager, California Department of Fish and Game, who provided breeding data on the osprey pair at Shellmaker Island; Scott Thomas and of the Audubon Society, who provided breeding data for both osprey pairs; Nancy Kenyon of the Audubon Society, who provided breeding data on the osprey pair at the San Joaquin Wildlife Sanctuary

Productivity can be expressed as fledglings per breeding attempt. With 23 fledglings and 11 breeding attempts, the productivity is 23/11 = 2.09 fledglings/breeding attempt. This high level of productivity of ospreys in Newport Bay suggests that DDT levels in eggs are below the threshold associated with hatching failure. Because ospreys are particularly sensitive to the most toxic effect of DDT, namely, on reproduction, they serve as an ongoing sentinel for the potential of DDT to affect all wildlife.

Discussion and conclusions. Considering the information on the effects of DDT on osprey reproduction that has been reviewed and summarized, can a threshold for the action of DDE on reproduction in osprey be determined? A lot is known. However, one is also aware of unknown exposures and the high variability of residues and response. A wide range of endpoints and approaches can be taken.

For a given breeding area, a field study, in which no significant eggshell thinning was found defines the known threshold for any biological effect of DDE residues in those eggs. The threshold for that finding is approximately several hundred ppb DDE. A threshold for increased shell breakage and reduced hatchability is approximately 3–4 ppm DDE. Recovery and stabilization of DDE-poisoned populations of ospreys has been associated with DDE egg residue levels as high as 5–8 ppm.

Although postulated, toxicity has not been shown to occur for DDE residue levels in eggs that cause shell thinning up to 10%. Thinking in evolutionary terms, normal eggshell thickness evolved to prevent breakage during incubation, as well as to provide gaseous exchange and an appropriate degree of hydration. There is a considerable range in normal eggshell thickness. Neither hatching success, nor health of the fledgling appears to be compromised by minimal shell thinning. There is some uncertainty here, but the recovery, stability and health of populations still experiencing marginal shell thinning, suggests no detrimental effect. In addition to choosing a threshold for toxicity, one must also determine an appropriate biomagnification factor from fish to egg.

Osprey are opportunistic feeders, catching the most nutritious and easiest to catch species at any given location and time. Typical prey species vary with season, latitude and whether the location is coastal or inland. One should expect, therefore, some variation in biomagnification from fish to eggs. The variation in literature biomagnification values, however, appears more related to a lack of representative sampling of fish from breeding grounds and a lack of data on residues in fish from wintering grounds.

The flounder, menhaden and largescale sucker appear to be the most important food species for osprey studied in North America. The largescale sucker is a fresh water species. Only the menhaden is among the species relied upon by the NAS panel in setting the marine fish recommendation to protect wildlife. For the 22 determinations of biomagnification from fish to egg determined from data in the reports above, there is considerable uncertainty. Therefore, the best estimate from these data for what constitutes an appropriate biomagnifications factor might be the median value of 10 (0.73–87, $n=22$). Values based on fish scraps cast from the nest range from 1.6 to 31 ($n=5$) with a median of 10.9. For reasons explained previously, a value of 10 is most likely to be high. For example, the two values from nonmigratory populations were 0.73 and 3.8.

A recommendation for DDT residues in marine fish should not consider DDD, because DDD has not been shown to cause shell thinning and is not converted to DDE. DDE causes eggshell thinning and DDT can be converted to DDE. Thus, DDT and DDE are the important terminal residues.

If the recommendation is to protect the osprey as a sensitive representative for other fish-eating species, then one needs to select a threshold level in eggs and divide by an appropriate biomagnification factor. If one were to use a threshold that is half of the approximate lower end of the hatchability effect threshold and divide that value by a biomagnification factor of 10, the recommendation would be 150 ppb in fish. This level is three times what the NAS panel recommended, but is based on additional information that they appeared to overlook or wasn't known until after 1972. The value of 150 ppb also served as the basis for the current National criterion for the water column discussed below.

3.2.3 US EPA Water Column Guidance to Protect Wildlife

In 1980, the US EPA published criteria for the protection of wildlife from DDT in the water column. The criterion was adopted as the California Toxics Rule (CTR) standard in 2002. The wildlife criterion of 1 pptr was based on the bioaccumulation of DDT from water into fish. A fish target residue was chosen to be 150 ppb from a study by Anderson et al. (1975) in a population of brown pelicans recovering from the reproductive effects of DDT. Accurate analysis of pptr levels of DDT in water is difficult and produces uncertain results, limiting the utility of the criterion. Measuring levels of DDT in fish is much easier and more certain. Therefore, a criterion in fish is more useful than a criterion in water. This brings us to the question of whether the 150 ppb DDT residue level in fish, which is the basis for the National criterion and CTR standard in water, will protect wildlife, considering what is known today. To remain consistent with the criterion, the fish residue should protect the brown pelican, one of the most sensitive species to the reproductive effects of DDT. This question is addressed in the following review of studies of DDT effects on reproduction in brown pelicans.

A summary of the 1980 US EPA criterion is reproduced below:

A residue value for wildlife protection of 0.0010 μg/l is obtained for both freshwater and saltwater using the lowest maximum permissible tissue concentration of 0.15 mg/kg based on reduced productivity of the brown pelican (Anderson, et al. 1975). Average lipid content of pelican diets is unavailable. Clupeids usually constitute the major prey of pelicans, and the percent lipid value of the clupeid, northern anchovy, is 8 (Reintjes, 1980). The northern anchovy is in some areas a major food source of the brown pelican. Therefore, the percent lipid value of 8 was used for the calculation of the Final Residue Value. The value of 0.15 mg/kg divided by the geometric mean of normalized BCF values (17,870) and by a percent lipid value of 8 gives a residue value of 0.0010 μg/l (Table 5).

Selection of the lowest freshwater and saltwater residue values from the above calculations gives a Freshwater Final Residue Value of 0.0010 μg/l and a Saltwater Final Residue Value of 0.0010 μg/l. The Final Residue Values may be too high because they are based on a concentration which reduced the productivity of the brown pelican.

The particular pelicans studied by Anderson et al. (1975) were reported to be feeding on northern anchovies. The northern anchovy diet of the recovering population of brown pelicans became the basis of the EPA chronic criterion to protect wildlife. The fish residue of 150 ppb is based on a study in which the brown pelican population was still recovering from the reproductive effects of DDT. The level of reproduction was judged to be inadequate to sustain the population. However, the authors referred to a slow reduction of DDE residues in brown pelican eggs compared to the fish diet. The fish residue had declined 27-fold during a period in which the egg residues had declined only ninefold. Moreover, DDT and DDE were detected in fish in 1974, but only the more stable DDE was detected in brown pelican eggs that year. Therefore, DDE in brown pelicans and in their eggs appears to have not reached a steady-state with the more rapidly declining residues in the aquatic environment. If we assume that in time the DDE residue in the eggs would also decline 27-fold, the final egg residue would have a geometric mean of 1.7 ppm. Would this level be a no-effect level for reproductive effects in the brown pelican?

To answer this question, let us review in chronological order some of the key studies of the effects of DDT on various populations of brown pelicans during and after the DDT era. The recovery of brown pelicans following the ban of DDT in 1972 provides a measure of dose-response and thresholds for the reproductive effects of DDT. Some of the earlier studies have been reviewed by Ware (1975).

Risebrough et al. (1967) reported the accumulation of DDT in higher trophic levels along the California coast. "Fish from California coastal waters contained more residue, but in general total concentrations were 10–20% of those in the birds." Bird species included Cassin's auklet, western gull, pelagic cormorant, Brandt's cormorant, brown pelican, common murre, ancient murrelet, red phalarope, rhinoceros auklet, sooty shearwater and slender-billed shearwater. Whole bird tissue contained from 1.0 to 15.4 ppm DDT. Western gull and Cassin's auklet eggs contained 6.5 ppm and 10.8 ppm, respectively. Fish included northern anchovy, English sole, Pacific jack mackerel, and hake. DDT levels in fish ranged from 0.2 to 2.8 ppm, with one sample of northern anchovy taken off Terminal Island, Los Angeles containing 12.7 ppm DDT.

In a 1969 conference at Oregon State University, Keith (1969) stated that scientists now have data to show that DDT is causing eggshell thinning in birds. Pelicans on Anacapa Island off the southern California coast produced good numbers of young in 1962, 1963 and 1966. In 1968 they were clearly in trouble, and in 1969 their reproductive effort was for all practical purposes a complete failure. In the same conference, Risebrough (Terriere et al. 1969) stated in a panel discussion that DDT levels in northern anchovies were low around San Francisco Bay compared to 5–15 ppm in waters off southern California. "We are aware of certain massive "hot spots": Clear Lake, California, Lake Michigan and evidently the Southern California coast." DDT stored in fat is toxicologically inert unless mobilized due to mobilization of fat stores. In a separate paper at the conference, Risebrough et al. (1969) spoke of recent findings, summarized as follows:

The DDT congener p,p'-DDE was the major cause of eggshell thinning in raptorial and fish-eating birds (Risebrough et al. 1969). The peregrine falcon, bald eagle and osprey were in decline due to DDE eggshell thinning. There was no evidence of thinning in eggshells of species that prey mostly on mammals, such as the Red-tailed hawk, golden eagle and great horned owl. Brown pelicans had declined 50% in the past 4 years at Point Reyes. Brown pelican and double-crested cormorant reproduction on the Channel Islands and Islas Coronados near San Diego were decimated in 1969. Western gull eggs on Anacapa Island in 1969 were normal. Some eggshell thinning is evident in ashy petrel and murre from the Farallon Islands. A "No effect" level has not been established for eggshell thinning. The relationship between DDE residues and eggshell thinning is linear with an absence of a "no effect" range of concentrations. DDE plus DDD in eggs from white pelicans, at levels ranging from less than 0.5 to 6 ppm, were associated with significant eggshell thinning.

"The complex series of behavioral events that lead up to mating, nest building, and egg laying were evidently not adversely affected." The likely mechanism of action is inhibition of calcium transport and mineralization in the shell gland. In the brown pelican, eggshell thickness is reduced about 15% at 75 ppm DDE on a lipid basis (3.3 ppm fresh weight). At higher residue levels the slope of the residue-thinning curve decreases to zero thickness at 3,000 ppm DDE (132 ppm fresh weight).

Keith et al. (1970) also studied the brown pelicans on the Channel Islands. Brown pelican eggshells from Anacapa Island were 34% thinner than pre-DDT era controls. DDE residues in the eggs were 29–183 ppm. DDE in brain tissue was high but not as high as the 30–60 ppm considered lethal.

Blus (1970) reported a study of eggshell thinning and breeding success in brown pelicans in Florida and South Carolina. Populations in both states were declining. Eggshells were 6–16% thinner than pre-DDT eggshells. Brown pelicans have been extirpated in Louisiana and other Gulf Coast localities. The reproductive failure and population declines were attributed to eggshell thinning caused by DDE.

Risebrough et al. (1971) reported an account of almost complete reproductive failure of brown pelicans on the Channel Islands in 1969. Broken and crushed eggs were strewn about the breeding area. Eggshell thickness was reduced 50%. Only two young were observed out of 1,272 nests.

A statistical analysis of the variability in eggshell thinning in brown pelicans implicated DDE as the causative organochlorine (Blus et al. 1971). Ten eggs from California contained DDE residues as high as 135 ppm with shell thinning of 25–35%. DDE residues in eggs from nine colonies in Florida ranged from 0.2 to 6.0 ppm. Eggs from two colonies in South Carolina had DDE residues ranging from 3.3 to 10.6 ppm. Blus et al. (1972a) reported that eggshell thinning of 15–20% had been associated with declining populations of several species of birds. The dose-response of DDE residue in eggs and eggshell thinning in brown pelicans was log-linear (Fig. 15). The estimated no-effect level was 0.5 ppm. The brown pelican is unusually sensitive to eggshell thinning by DDE. Fifteen percent thinning occurs at 4–5 ppm DDE in eggs. The herring gull showed no thinning when DDE residues in eggs were 4–5 ppm. The level of DDE in eggs is taken as an indication of DDE residues in the female.

The paper by Blus et al. (1972a) was accompanied by a letter from William Hazeltine challenging the assertion that the DDE eggshell thinning dose-response was log-linear. Moreover, Hazeltine questioned whether DDE caused eggshell thinning. He suggested scientists were acting irresponsibly to ban pesticides.

Risebrough (1972) also wrote a letter to the same journal. His letter defended Blus et al. and refuted Hazeltine's comments. He stated that in some cases the log-normal distribution provides an excellent fit to the brown pelican data, and "In several other cases the gamma distribution more adequately describes the observed distribution of pollutants."

Switzer et al. (1972) also wrote a letter challenging Blus et al.'s conclusion that eggshell thinning in the brown pelican was caused by DDE. They pointed out that museum eggs, used to establish pre-DDT era shell thickness, were often selected as the best (and perhaps thickest) specimens for display in public exhibits.

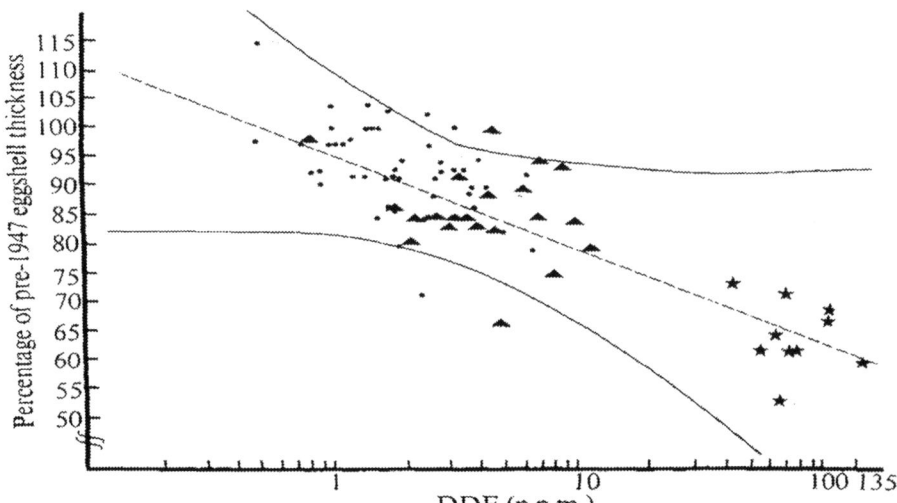

Fig. 1 Association of DDE residues in eighty brown pelican eggs from Florida (●), South Carolina (▲) and California (★) with the % of pre-1947 eggshell thickness. Solid lines represent 95% confidence limits. $Y = 95.787 - 15.689 \log_{10} X$; $r = -0.80$ ($P < 0.01$).

Fig. 15 Relationship of DDE in eggs and eggshell thinning in brown pelicans. Figure 1 in Blus et al. (1972a) reproduced with permission from the author and from Nature

Blus et al. (1972b) responded to comments by Hazeltine and by Switzer et al. in a follow-up report. They pointed out that lipid levels in eggs decrease about one-third from laying to hatching. Since DDE residues are localized in the lipid, the lipid concentration of DDE will increase during incubation.

Schreiber and Risebrough (1972) published a review of the status of the brown pelican in the United States and Baja, Mexico. They also reported on Schreiber's work on brown pelicans in Florida. Hatching success in Florida decreased sharply with increasing frequency of inspection by wildlife biologists. The lipid content of Florida eggs was 5.0%. The authors claimed that very low concentrations of DDE were associated with significant thinning and that the relationship is linear from zero concentrations of DDE. Thinning of eggshells greater than 20% usually causes them to break during incubation. Total DDT residues in eggs collected in 1969 and 1970 in Florida were 1.2–2.9 ppm. The 9% reduction in eggshell thickness in Florida had not yet had an observable effect on population stability. There was no evidence that 9% shell thinning has an effect on gas exchange or water retention.

Keith and Gruchy (1972) published a comprehensive review of the past 5 years of reports on the effects of DDE on avian wildlife. They noted a wide species variation in eggshell thinning response to DDE residues (Fig. 16).

Jehl (1973) reported on the status of brown pelicans on islands off the west coast of Baja, California. Breeding was severely impacted at most of the locations, with empty nests and broken shells. Observations were complicated by destruction of

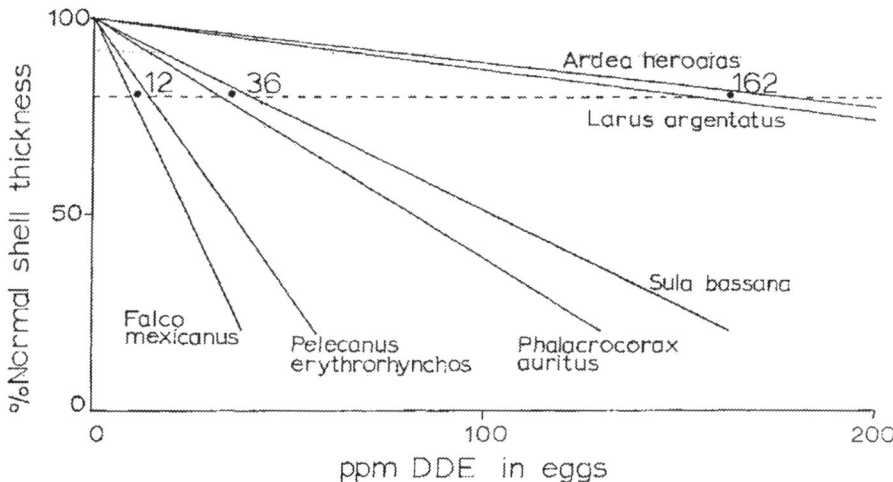

FIGURE 7. Variation between species in reproductive response to DDE. DDE values are of whole eggs on a wet-weight basis. Twenty-percent reduction in shell thickness (population damage threshold) is shown by a dotted line, and the calculated DDE values at that thickness are shown as 12, 36, and 162 ppm for the three pairs of slopes. Sources are *Pelecanus erythrorhynchos* and *Phalacrocorax auritus*, [1]; *Falco mexicanus*, [11]; *Larus argentatus*, [14]; and *Sula bassana*, unpublished data of J. A. KEITH.

Fig. 16 Species variation in eggshell thinning in response to DDE egg residues. Figure 7 in Keith and Gruchy (1972) reproduced with permission

pelican eggs by gulls whenever nests were unattended. The source of DDE was attributed to the Los Angeles outfall. The dose-response for DDE in eggs and shell thinning is presented in Fig. 17 (DDE concentration in ppm lipid).

Blus et al. (1974a) reported on studies of brown pelican eggs collected in 1969 and 1970 from California, Florida and South Carolina. Eggshells were thinner than pre-DDT era eggshells. DDE residues were highest in California eggs and lowest in Florida eggs. Shell thinning was highly correlated with levels of DDE in the eggs. The calculated no-effect level was 500 ppb DDE. Thinning was 4% at 1 ppm and 15% at 5 ppm. The observed logarithmic relationship was also reported by others for the double-crested cormorant and the prairie falcon. Dieldrin may have contributed to reproductive failure of brown pelicans. Serious population declines have occurred in California and South Carolina as a result of DDE eggshell thinning. "The 17% eggshell thinning observed in South Carolina was associated with subnormal reproductive success." In areas with the greatest eggshell thinning, "Usually, the entire clutch exhibited the extreme thinning, and all the eggs were broken in some nests." Florida eggs from different breeding areas averaged 0.69–2.48 ppm DDE, with an average of 8% shell thinning. "…the bulk of the residues in all areas of Florida are low enough that one would not expect these residues to induce widespread, long-term, adverse effects on the populations there." The log-linear relationship between DDE residues in eggs and shell thinning is illustrated in Fig. 18 below.

A more systematic study was done in 1971 and 1972 by Blus et al. (1974b) in a breeding colony of brown pelicans in South Carolina. One freshly laid egg was

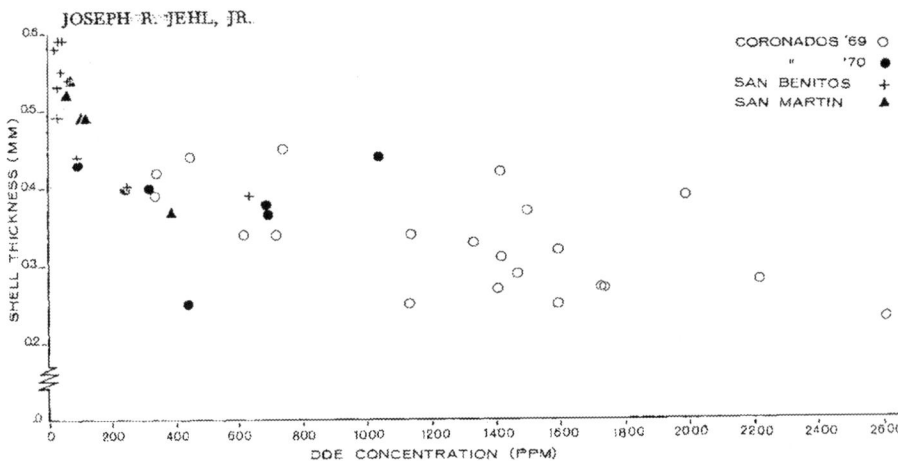

FIGURE 3. Relationship between DDE concentration and shell thickness in Brown Pelican eggs from northwestern Baja California.

Fig. 17 DDE dose-response in eggshell thinning in brown pelicans on the islands off the west coast of Baja California. Figure 3 in Jehl (1973) reproduced with the permission from the author and from The Condor, published by The Ornithological Society

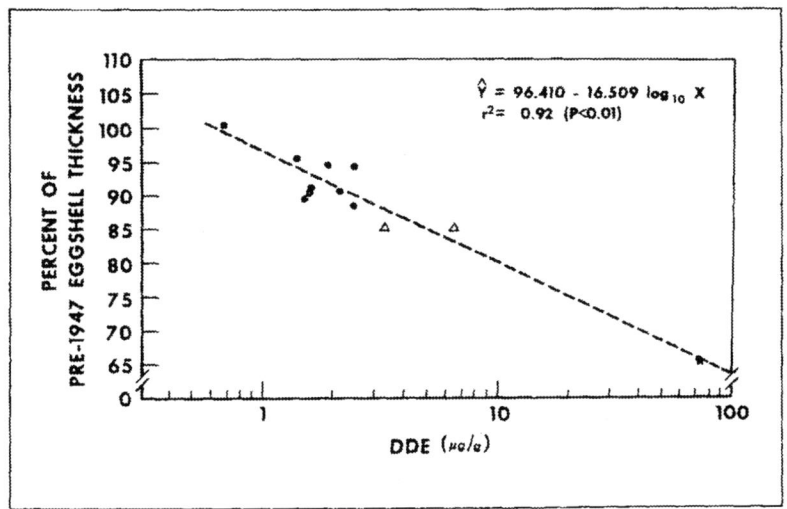

FIGURE 2.—*Association of DDE residues in brown pelican eggs from nine collections in Florida [●], two colonies in South Carolina [△], and one colony in California [★] with percent of pre-1947 eggshell thickness.*

Fig. 18 Log-linear relationship between DDE residues in eggs and eggshell thinning in brown pelicans from Florida, South Carolina and California. Figure 2 reproduced from Blus et al. (1974a)

Table 23 Early recovery of brown pelican reproduction off Baja and California coasts. Table 1 in Anderson et al. (1975) reproduced with permission from the author and from AAAS

Table 1. Recent history of brown pelicans breeding off the coast of southern California and northwestern Baja California; productivity totals include Anacapa and Santa Cruz Islands and Isle Coronado Norte (3). Abbreviation: C.L., confidence level.

	No. nests built	No. young fledged		Eggshell thickness*				Reference	Anchovy abundance†
				Crushed/broken		Found intact			
		Total	Per nest	No.	$\bar{X} \pm 95\%$ C.L. (mm)	No.	$\bar{X} \pm 95\%$ C.L. (mm)		
1969	1125	4	0.004	53	0.288 ± 0.016	12	0.402 ± 0.019	(14)	140
1970	727	5	0.007	72	0.286 ± 0.014	16	0.393 ± 0.021	(28)	70
1971	650	42	0.065	17	0.310 ± 0.030	6	0.460 ± 0.026		80
1972	511	207	0.405	25	0.294 ± 0.034	4	0.438 ± 0.024		195
1973	597	134	0.225	26	0.343 ± 0.033	4	0.510 ± 0.068		225
1974	1286	1185	0.922	27	0.378 ± 0.033	59	0.482 ± 0.016		355

*Arithmetic means are given. Normal eggshell thickness for this population is 0.572 ± 0.010 mm ($N = 11$) (9); eggshells were measured by standard techniques (9). Intact eggs included some destroyed by predators. Thickness data for 1969 to 1973 are from Anacapa and Santa Cruz only; those for 1974 also include samples from Isla Coronado Norte, which were not significantly different. † This is an estimate of biomass expressed as thousands of schools per census in a fixed area off southern California during January to June, as derived from figure 6 of Mais (4).

taken from each of 93 marked nests. In this way, residue level and shell thinning could be related directly to nest success. The effects of DDE on eggshell thinning and reproductive success were confounded by dieldrin. Reproductive success was normal in those nests in which a sample egg contained less than 2.5 ppm DDE.

Anderson et al. (1975) published the first report of the recovery of the brown pelican following the ban of DDT in 1972. The major source of DDT for the study populations was the wastes of the DDT manufacturer being released into the ocean by way of the Los Angeles County Sanitation District's waste water outfall at White's Point. Releases were greatly reduced after April, 1970. Recovery of brown pelican reproduction on offshore islands to the north and south improved quickly during the period 1971–1974 (Table 23).

Fledging rates increased from 0.004 to 0.922. Thicker shelled eggs and fewer broken eggs were observed over time during this period. The recovery was not complete, as a fledging rate of 1.2–1.5 is needed to achieve a stable population.

Direct observation confirmed that the northern anchovy was the major food item for this breeding colony of brown pelicans. The authors (Anderson et al. 1975) stated: "During banding at Anacapa from 1972 to 1974, we examined stomach contents regurgitated by young pelicans; the material consisted almost exclusively of anchovies. Our observations of feeding adults before and during the breeding season also indicated a heavy reliance on anchovies."

Residues of DDE in northern anchovies decreased 27-fold from 1969 to 1974 while during the same period, DDE in brown pelican eggs decreased ninefold (Table 24). The slower decline in egg compared to fish residues suggests that at a steady-state, the 150 ppb total DDT measured in northern anchovies in 1974 would result in an egg residue that is below the threshold for a reproductive effect.

Anderson et al. (1977) continued to study brown pelicans on Anacapa Island in 1975. The only breeding colonies in California observed by these investigators were on Anacapa Island and nearby scorpion rock. Only four eggs were collected and three of these were putrified. Lipid content of eggs was assumed to be 5%. DDE egg residue analysis, shell thickness and productivity appeared to have leveled off in 1975, following the recovery from 1969 to 1974. PCB egg residues were 5–10 ppm during this period. The data are summarized in Fig. 19. PCBs may have affected reproduction,

Table 24 Decreases in DDE in northern anchovies and brown pelican eggs off the Southern California and Northwestern Baja California coasts during 1969–1974. Table 2 in Anderson et al. (1975) reproduced with permission from the author and from AAAS

Table 2. Geometric mean residues of DDT and related compounds (DDE and TDE) (12) in anchovies and brown pelican eggs off the southern California and Baja California coasts. Abbreviations: Cr, crushed eggs; In, intact eggs; N.D., residues were not detected (< 2 ppm, lipid basis) (24).

Year	Anchovy whole bodies*				Brown pelican egg contents[†]				Reference
	Residue (ppm, fresh weight basis)				Residue (ppm, lipid weight basis)				
	No.	DDT plus TDE	DDE	Total	No.	DDT plus TDE	DDE	Total	
	Southern California and northwestern Baja California								
1969	11	1.03	3.24	4.27	73 (Cr)	49.0	1155.3	1204.3	(14)
					28 (In)	54.2	852.5	906.7	(29)
1970	15	0.56	0.84	1.40					
1971	6	0.47	0.87	1.34					
1972	8	0.38	0.74	1.12	10 (In)		220.9	> 220.9	
1973	4	0.11	0.18	0.29	4 (In)	6.5	174.9	182.9	
1974	4	0.03	0.12	0.15	39 (In)	N.D.	96.6	96.6	
	West-central Baja California								
1969	10	0.06	0.20	0.26	16 (In)	5.8	89.5	96.1	(14)

*Anchovies were collected from January to August each year. Individual fish were analyzed in 1969 and pools of 10 to 30 fish were analyzed thereafter; sensitivity was 0.01 ppm (24). The anchovies from west-central Baja California probably represent a different population (5).   †Eggs from Coronado Norte were included only in 1969 and 1974. The pelican eggs from west-central Baja California were collected at Isla San Benito.

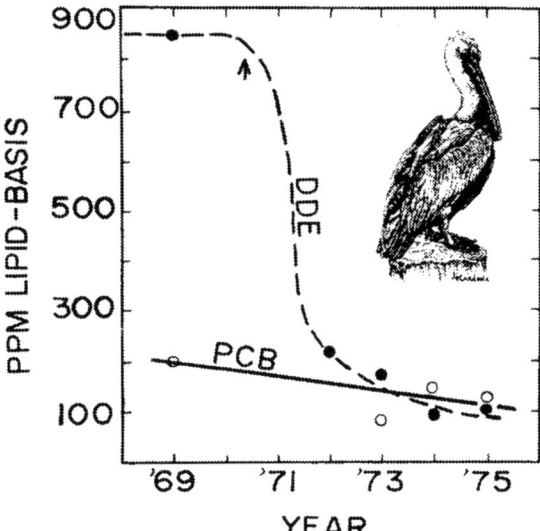

FIGURE 1. Residue changes of DDE and PCB in intact brown pelican eggs from Southern California. The arrow indicates a major drop in environmental input of DDT. According to published data, major input of DDT ceased in 1970 (Anderson et al. 1975) and by 1971 it had decreased to about 0.5% of previous levels (Jukes 1974, citing the DDT-manufacturing company president). There is some disagreement as to the actual levels of input before 1970 (Jukes 1974).

Fig. 19 DDE levels in brown pelican eggs from Southern California. Figure 1 in Anderson et al. (1977) courtesy of the author and California Fish and Game

Table 25 DDT and metabolites in Atlantic menhaden regurgitated by brown pelicans in South Carolina in 1973. Data from Table 14 in Blus et al. 1977

Residues (µg/g fresh wet wt)		
DDE	TDE(DDD)	DDT
0.04	0.04	0.04
0.06	0.04	0.05
0.07	0.02	0.03
0.06	0.03	0.03
0.05	0.03	0.02
0.08	0.03	0.02
0.15	0.07	0.06
GM[a] 0.067	0.035	0.033
CL[b] 0.045–0.099	0.024–0.050	0.022–0.049
Range 0.04–0.15	0.02–0.07	0.02–0.06

[a]GM = geometric mean
[b]CL = 95% confidence limits

preventing the full recovery of the colony. Limited observations in 1976 suggested that an inadequate food supply also contributed to low productivity.

In 1977, Blus et al. published a follow-up report on the brown pelican breeding colonies in South Carolina. Shells of eggs collected from 1969 to 1973 averaged 14–17% thinner than shells of eggs collected prior to the DDT era. Crushed shells were thinner than shells from eggs that hatched. Shells of freshly laid eggs were thinner than shells of hatched eggs. Residues of DDE in eggs decreased from 5.45 ppm in 1969 to 2.09 ppm in 1973. Reproductive success of 1.66 per nest in 1973 was considered excellent. Atlantic menhaden, a major food item of the brown pelican, contained a residue of 0.135 ppm total DDT as shown in the author's Table 14 that is reproduced in Table 25 below. The menhaden were recovered from regurgitated stomach contents in 1973. The biomagnification value for total DDT from fish to egg was 18. The total DDT residue in menhaden in the late 1960s was 0.295 ppm. "The migratory habits of the Atlantic menhaden (15, 17) and the brown pelican confound the significance of biomagnification noted in this study."

Thompson et al. (1977) reported on a 1970–1971 study of brown pelicans in Florida. Regurgitated food items from 14 colony sites were analyzed and found to contain an average of 0.074 ppm total DDT in 1970 and 0.047 ppm in 1971. Total DDT in fish collected in 1964–1965 averaged 0.174 ppm. Total DDT in brown pelican eggs collected in 1971 from three colony sites averaged 1.27 ppm.

King et al. (1978) reported on DDT residues and shell thinning in addled brown pelican eggs collected in 1970 along the Texas coast. The average total DDT residue was 3.23 ppm and was negatively correlated with an average 11% shell thinning.

King et al. (1977) reported 10% thinning in brown pelican eggs collected in Texas from 1970 to 1974. DDE levels declined from 3.2 ppm in 1970 to 0.86 ppm in 1974. Endrin toxicity accounted for mortality in adult pelicans and may have caused reproductive failure. Effects of DDE on reproduction during this period could not be assessed due to the small populations and confounding endrin toxicity.

Mendenhall and Prouty (1978) studied recovering populations of brown pelicans in South Carolina. A steady decline in DDE residues in eggs had a high negative correlation with increasing eggshell thickness (Fig. 20). Eggshell thickness in 1978 was only 6% below the pre-1947 mean thickness. Fledgling rates continued to increase and reached a population sustaining level in 1976 (Table 26). The authors noted that in 1977 all eggs sampled were below 2.5 ppm DDE. DDE levels above 2.5 ppm had been associated with consistent nest failure.

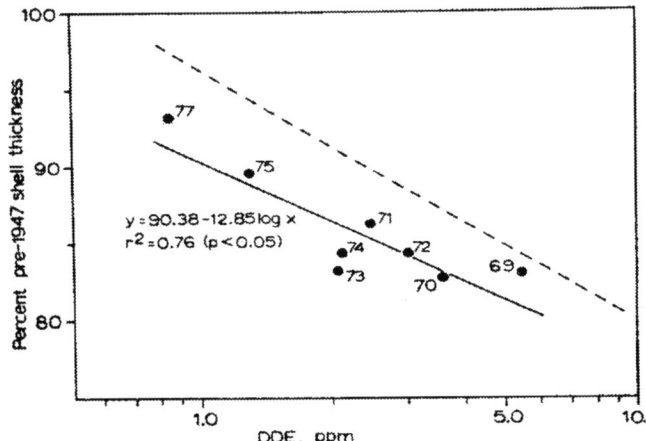

Fig. 1. (———) Change in eggshell thickness for South Carolina brown pelicans as related to DDE residues, 1969-1977. Each point shows mean shell thinning in relation to pre-1947 data (y) and mean wet-weight DDE residue (x) for one year. Sources of data as in Table 2. (----) Regression for 12 colonies in 3 states, 1969-70; $y = 96.410 - 16.509 \log_{10} x$, $r^2 = 0.92$ (Blus et al. 1974a).

Fig. 20 Negative correlation between egg DDE residue and shell thickness for South Carolina brown pelicans. Figure 1 in Mendenhall and Prouty (1978) reproduced with permission

Table 26 Recovery of reproductive success in brown pelicans from South Carolina from 1969 to 1978. Table 3 in Mendenhall and Prouty (1978) reproduced with permission

TABLE 3.
Colony size (peak nest numbers) and fledging success for Brown Pelicans, South Carolina, 1969-1978.

	1969	1970	1971	1972	1973	1974	1975	1976	1977	1978
Nests	1266	1116	1469	1415	1646	1670	2400	2540	3376	3353
Fledged per nest	0.78	0.85	0.92	0.69	1.66	0.97	0.75	1.23	1.4*	1.35

* Approximate figure; see text.

Sources: 1969-76, as Table 2; 1977-78, present study.

Blus et al. (1979a) reported on a program to transplant brown pelicans from Florida to Louisiana. A total of 765 young pelicans were transplanted in 1971 and began breeding and increasing in numbers until a severe die-off in 1975. The die-off was attributed to endrin. Eggshell thickness gradually decreased to 14% below pre-DDT era thickness by 1974 and then began to increase thereafter. Endrin use was curtailed in 1976 and breeding improved to 1.47 fledged per nest. The authors considered fledgling rates of 1.2–1.5 to be necessary to maintain a stable population. DDE residues in eggs peaked at 1.36 ppm in 1972 and decreased to 0.92 ppm by 1976. The authors concluded that DDE-induced eggshell thinning was not high enough to interfere with reproductive success.

Blus et al. (1979b) reported on DDT residues, eggshell thinning and reproduction in brown pelicans in South Carolina and Florida. Samples of the primary food item of the breeding colonies, the Atlantic menhaden, were collected in 1974 and 1975 from regurgitated stomach contents in South Carolina and analyzed for DDT. From 1969 to 1975, the trend in total DDT residues in eggs from South Carolina was steadily downward from 7.81 to 1.80 ppm. DDE decreased from 5.45 to 1.40 ppm during the same period. By 1975, residues of parent DDT were barely measureable. Menhaden DDE residues were 0.016 ppm in 1974 and 0.014 ppm in 1975. Egg shells increased in thickness from 17% thinner to 10% thinner than pre-DDT era eggshells. Florida populations had been stable for several years. South Carolina populations were increasing. Fledgling rates in the South Carolina populations in 1975 were adequate to maintain a stable population.

Blus (1982) provided further interpretation of the relationship of DDT residues in brown pelican eggs to reproductive success. By collecting single eggs from a marked nest and following productivity in the same nest, residues of DDE could be associated directly with reproductive success. The critical level of DDE residues in eggs was 3 ppm. Residues below this level generally produced, at most, a slight reproductive effect. Residues in excess of this level were associated with a substantial effect on reproduction. A residue of 4 ppm in eggs was associated with total reproductive failure.

An overall decline in organochlorine residues in brown pelican eggs is illustrated in data from Blus et al. (1979b), which is plotted in Fig. 21 below.

In 1983, Anderson and Gress published an update on the status of populations of brown pelicans in the Southern California Bight. DDE residues in eggs and eggshell thinning were not measured. Fledgling rates were closely associated with stocks of northern anchovies since about 1974. The population of brown pelicans on Anacapa Island continued to increase even though fledgling rates were below one; "…1980 was the first year when reproduction was probably not drastically affected by pollution…".

Blus (1984) reported a comparison of regression and sample egg methods for predicting the reproductive effects threshold for DDE. Brown pelican eggs from California, Florida, Louisiana, and South Carolina were analyzed for DDE residue, eggshell thinning, and these were then compared to reproductive success. Eggshell

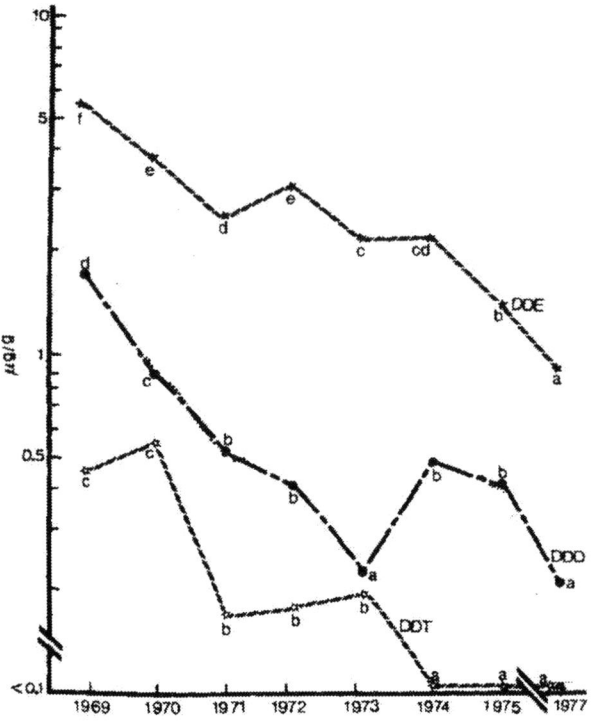

Fig. 21 Decline in DDT and metabolites in eggs of brown pelicans collected in South Carolina from 1969 to 1977. Data from Figure 1 in Blus et al. (1979b)

thinning of 18% or greater had been reported to be associated with reproductive failure and population declines. An DDE egg residue of 5 ppm was associated with 18% shell thinning by regression analysis (Fig. 22). Using the sample egg method, reproductive effects occurred at 3 ppm, with a threshold between 2.5 and 3 ppm DDE (Fig. 23). The critical level of 3 ppm is associated with eggshell thinning of 16% from the regression analysis of 813 eggs plotted in Blus's Figure 1.

In 1985, King et al. reported on studies from 1975 to 1981 on colonies of brown pelicans in Texas. During this period, nesting pairs increased from 18 to 57. Fledgling rates were considered adequate in all years except 1975. DDE levels were about half those measured in 1970 and ranged from 0.9 to 2.3 ppm. "Current levels of DDE apparently pose a minimal threat to pelican reproduction." "Mean eggshell thickness was 4–14% thinner than normal, but we found no evidence that shell thinning adversely affected reproduction." DDE residues in a major food item, the gulf menhaden, were measured at an average of 0.06 ppm in 11 fish in 1980. "DDT and metabolite residues may have been magnified 23 times from fish (0.06 ppm) to pelican eggs (1.36 ppm), but interpretation of this apparent biomagnification is complicated by the migratory habits of the pelicans and their prey."

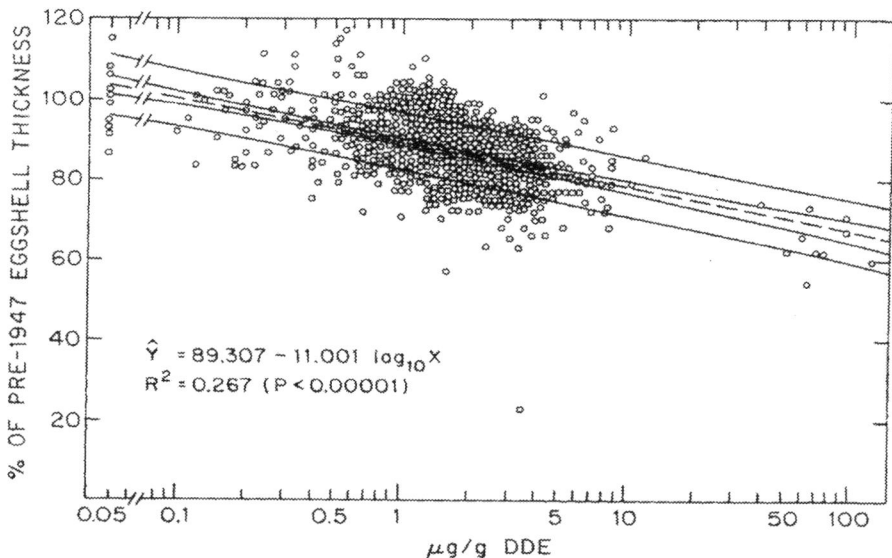

FIG. 1. Regression analysis showing the relationship of DDE residues in 813 eggs of Brown Pelicans to eggshell thickness; South Carolina, Florida, Louisiana, and California, 1969–1976. The dashed line is the regression line, the two pairs of solid lines delineate the 95% confidence limits for the population mean (inner pair) and for individual eggs (outer pair).

Fig. 22 Regression of DDE residue and shell thickness in brown pelican eggs collected from 1969 to 1976. Figure 1 in Blus (1984) reproduced with permission from the author and from The Wilson Journal of Ornithology

Gamble et al. (1987) reported on a 1986 study of a colony of brown pelicans in Texas and two colonies in the Yucatan Peninsula in Mexico. DDE residues in eggs from Texas averaged 0.16 ppm. These levels reflected a tenfold decline from 1975 levels. The authors concluded: "The concentrations of the organochlorine compounds in eggs from Texas and Mexico were below levels considered to be harmful."

In 1995, Franklin Gress published his doctoral thesis on 22 years of studies of brown pelicans on Anacapa Island. DDE residues in eggs declined slowly during the late 1970s and 1980s to approximately 2 ppm in 1992. Eggshells increased in thickness during this period. Thinning was about 5% in 1992. Gress concluded: "... at present we have no evidence that brown pelican reproduction in the SCB is measurably impaired by DDE-related eggshell changes..." The only breeding colonies in the Southern California Bight (SCB) are on West Anacapa Island, Santa Barbara Island and Islas Los Coronados.

Discussion and conclusions. To summarize the key findings in the above chronology, brown pelican reproduction was reduced by the direct action of the DDT metabolite, DDE, during and after the DDT use era. DDE was magnified up the aquatic food

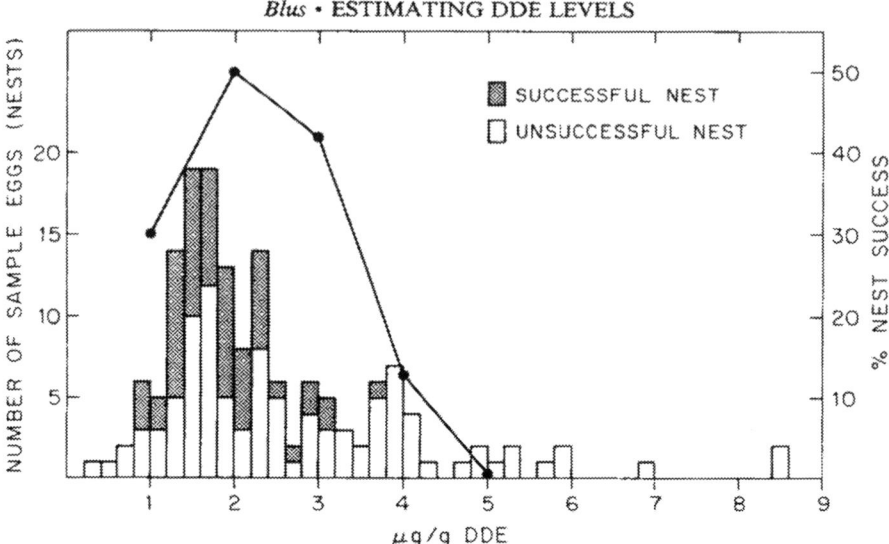

FIG. 2. Relationship of DDE residues in 156 sample eggs of Brown Pelicans to nest success. Bars represent success related to 0.2 µg/g intervals; dots on the line represent mean nest success by µg/g intervals.

Fig. 23 Threshold for DDE reproductive effects in brown pelicans. Figure 2 in Blus (1984) reproduced with permission from the author and from The Wilson Journal of Ornithology

chain to the fish diet of the brown pelican and was deposited in the lipid of the eggs. DDE residues exceeding 2.5 ppm in eggs were associated with eggshell thinning in excess of 15%, resulting in decreased hatching success. DDE egg residues below 2.5 ppm, although capable of producing measureable thinning of eggshells, were not associated with reduced hatching success or any other effect on reproduction. DDE residues in all populations of brown pelicans in the United States are currently below the threshold for reduced hatching success.

Brown pelicans in the Southern California Bight were most impacted by DDE during the 1960s and 1970s. The reason is the much higher contamination levels from the production wastes of DDT manufacture, compared to agricultural residues generated throughout the regions populated by brown pelicans. The highly contaminated Palos Verdes Shelf provides a continuing source of DDE to the northern anchovy diet of the breeding colonies of brown pelicans on Anacapa Island. For example, the Southern California Bight study of 1998 (Allen et al. 2002) found total DDT levels as high as 10.5 ppm in fish captured in the Palos Verdes Shelf area. This aquatic food-chain source explains the slow decline and leveling off of DDE residues in eggs collected on Anacapa Island. Breeding colonies further south, off Baja California, have much lower egg residues.

In spite of the high DDE levels on the nearby Palos Verdes Shelf, the brown pelicans on Anacapa Island apparently now have residue levels below the threshold for reproductive effects (Gress 1995). The steady-state residue level of 1.7 ppm DDE in eggs, estimated from the 1974 data, is below the threshold for reproductive effects based on the above review. This level is very close to what was measured in eggs from Anacapa in 1992.

Reports of DDE residues in the northern anchovy diet of brown pelicans were not found in published literature after 1975. Therefore, a confirmation of the biomagnification from fish diet to eggs of approximately 11, estimated from the Anderson et al. (1975) 1969 data from intact eggs, is not available.

However, one can conclude that the Anacapa breeding colony most likely represents a worst case for all other regions that are not directly influenced by DDT production wastes. That is, if reproduction in the Anacapa population is no longer affected by DDT, then one should expect that aquatic environments contaminated from agricultural use, a much lower level of contamination than that on the Palos Verdes shelf, are no longer at DDE levels that would affect reproduction in brown pelicans. In fact, the margin of safety for agricultural residues should be greater than that for the industrial wastes contaminating the food supply of the Anacapa colony.

3.2.4 Canadian Fish Guidance to Protect Wildlife

In 2000, Environment Canada published Environmental Quality Assessments for PCBs, DDT and Toxaphene (Environment Canada 2000). The Assessment document contains the derivation of a Canadian tissue residue guideline (TRG) for total DDT. The TRG for fish was intended to protect avian species from the reproductive effects of DDE. The TRG is based on low-observed-effect-levels (LOELs) for shell thinning in mallard and black ducks. Several generic assumptions were made to arrive at the TRG of 14 ppb in fish as shown in the text of the Assessments document as follows:

For birds exposed to DDT, the most sensitive endpoint appears to be eggshell thinning and associated reproductive impairment. The most sensitive LOAEL determined from the avian dataset was 0.3 mg·kg^{-1} bw·day^{-1}. The same LOAEL was determined from several studies on mallard ducks and black ducks. Eggshell thinning occurred when mallard ducks were fed 0.3 mg·kg^{-1} bw·day^{-1} of *p,p'*-DDT for 30 days (Kolaja 1977), 0.3 mg·kg^{-1} bw·day^{-1} of *p,p'*-DDE for 105 days (Vangilder and Peterle 1980), for 30 days (Kolaja 1977), and for 365 days (Heath *et al.* 1969). Black ducks showed a reduction in eggshell thickness when administered 0.3 mg·kg^{-1} bw·day^{-1} of *p,p'*-DDE for 136 days (Loncore *et al.* 1971). The NOAEL was assumed to be 0 mg·kg^{-1} bw·day^{-1}. For the purpose of calculating the TDI, the LOAEL was divided by 5.6 (according to CCME 1993) to estimate a NOAEL of 0.054 mg·kg^{-1} bw·day^{-1}.

According to Sample *et al.* (1996), avian studies where exposure duration is 10 weeks or less are considered to be sub-chronic, and those where the exposure duration is greater than 10 weeks are considered chronic studies. Several studies on the reproductive effects of DDT in birds were carried out for longer than 10 weeks, therefore these studies were considered to be chronic. Although no data were located on the carcinogenic or mutagenic effects of DDT in avian species, a large quantity of data exists on the effects of DDT to several avian species, including those known to be sensitive to the reproductive effects of DDT such as raptors. Therefore, an UF of 10 (CCME 1997) was used to account for differences in interspecies sensitivities. The LOAEL of 0.30 mg·kg^{-1} bw·day^{-1} was used in conjunction with the NOAEL of 0.054 mg·kg^{-1} bw·day^{-1} to calculate an avian TDI of 13.0 μg·kg^{-1} bw·day^{-1} for total DDT.

$$\text{TDI} = (0.30 \text{ mg·kg}^{-1} \text{ bw·day}^{-1} \cdot 0.054 \text{ mg·kg}^{-1} \text{ bw·day}^{-1})^{0.5} \div 10$$
$$\text{TDI} = 0.013 \text{ mg·kg}^{-1} \text{ bw·day}^{-1} = 13.0 \text{ μg·kg}^{-1} \text{ bw·day}^{-1}$$

The mammalian and avian TDIs were then used in conjunction with the body weights (BW) and daily food intake rates (FI) of the wildlife species with the highest FI:BW ratios to calculate reference concentrations (RCs) of total DDT, using the following equation:

$$\text{RC} = \text{TDI} \cdot (\text{BW} \div \text{FI})$$

where:
RC = Reference concentration (mg·kg^{-1} ww);
TDI = Tolerable daily intake (mg·kg^{-1} bw·day^{-1});
BW = Body weight (kg ww);
FI = Food intake rate (kg ww·day^{-1})

Among mammalian and avian wildlife species, female mink (*Mustela vison*) and Wilson's storm-petrel (*Oceanites oceanicus*) have the highest potential exposure to DDT due to their high FI:BW ratios (0.24 and 0.94, respectively) (CCME 1997). Therefore, these species were used to calculate the RCs for total DDT.

DDT

Similarly, a RC of 14.0 µg·kg^{-1} was calculated for Wilson's storm-petrel, assuming a body weight of 0.032 kg, an average daily food intake rate of 0.03 kg ww·day^{-1}, and a TDI of 13.0 µg·kg^{-1} bw·day^{-1} for birds (Dunning 1993).

$$RC = 13.0 \text{ µg·kg}^{-1} \text{ bw·day}^{-1} \cdot (0.032 \text{ kg} \div 0.030 \text{ kg ww·day}^{-1})$$
$$RC = 14.0 \text{ µg·kg}^{-1}$$

The lower of the mammalian and avian RCs, 14.0 µg·kg^{-1} was recommended as the Canadian TRG for total DDT for the protection of freshwater, marine, and estuarine wildlife that consume aquatic biota.

These assumptions and the procedures for deriving the TRG for DDT in fish were from a protocol document published by the Canadian Council of Ministers of the Environment (1999). The Protocol document calls for the use of "...sensitive endpoints, such as embryonic development, early survival, growth, reproduction, adult survival, and other ecologically relevant responses." This Protocol document states that an uncertainty factor of at least 10 is to be used to account for variability in species, gender, life stage, and duration of exposure. The Protocol document also recommends the use of a factor of 5.6 to extrapolate from a LOEL to a no-observable-effect-level (NOEL), if a NOEL cannot be estimated directly from dose-response data. Finally, TRGs are to be corrected for the species with the highest food consumption per body mass.

Environment Canada chose to use ducks as the test species and egg shell thinning as the toxic endpoint for assessing the reproductive effect of DDT on fish-eating avian species. Mallard and black ducks are primarily herbivores. They are also not particularly sensitive to the reproductive effects of DDE (Peakall et al. 1973; Peakall 1975). Eggshell thinning below the threshold for hatching failure has been shown in numerous studies not to be detrimental to avian wildlife. Environment Canada cites, but does not use, studies done with American kestrels (sparrow hawks). This hawk species is not fish-eating, but does feed on insects and small mammals. Laboratory and field studies have established a dose-response in eggshell thinning, DDE residues in eggs, and hatching failure (Porter and Wiemeyer 1969; Wiemeyer and Porter 1970; Peakall et al. 1973). Studies reported by Lincer (1975) contain concurrent laboratory and field studies. Residues in diet, eggs and eggshell thinning were used to correlate the field and laboratory studies. One can see a clear dose-response between shell thickness and DDE egg residue level (dry weight basis) using the combined laboratory and field data (Fig. 24).

The same data are summarized in Appendix 16 of the Assessment document (Environment Canada 2000) reproduced in Table 27 below. The 0.5 mg/kg-day level (3 ppm in the diet) produced 15% eggshell thinning, corresponding to a level just below the threshold for hatching failure, the most sensitive toxic endpoint of

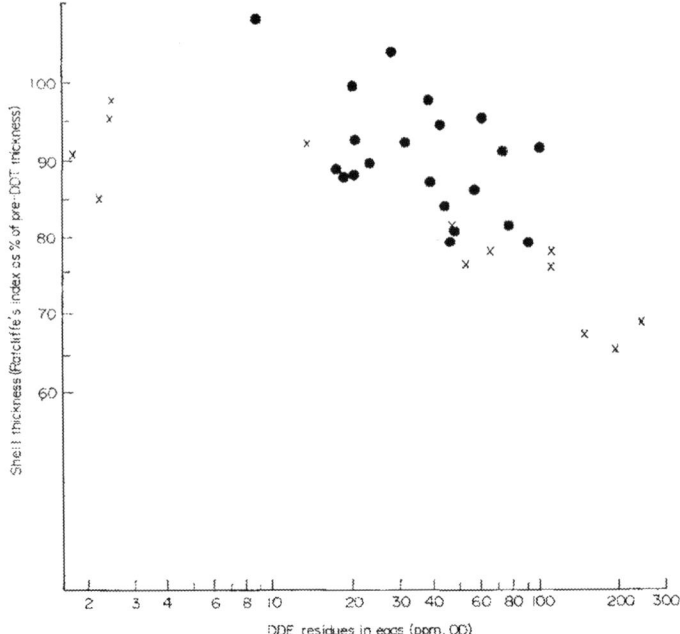

Fig. 3. Relationship between mean clutch shell-thickness and DDE residue of kestrel eggs collected in Ithaca, New York during 1970 (●) and same relationship experimentally induced with dietary DDE (×).

Fig. 24 Influence of laboratory and field dietary exposures to DDE on shell thickness in kestrel eggs collected in Ithaca, New York. Figure 3 in Lincer (1975) reproduced with permission from the author and from John Wiley & Sons

Table 27 Effect of dietary DDT on eggshell thickness in birds. Appendix 16 reproduced from Environment Canada (2000)

Appendix 16. Summary of data on the reproductive effects of orally-administered DDT and its metabolites on birds.

Species	Life Stage	Sex	Daily Dose (mg/kg BW/day)	Duration of Exposure (d)	Total Dose (mg/kg BW)	Endpoint Measured	Reference
p,p'-DDE (cont.)							
Black ducks	adult	F	0	136	0	Eggshell thickness (0.34 mm) - control	Longcore et al. 1971
Black ducks	adult	F	0.3	136	41	Eggshell thickness (0.28 mm) - S	Longcore et al. 1971
Black ducks	adult	F	0.9	136	122	Eggshell thickness (0.26 mm) - S	Longcore et al. 1971
Indian runner ducks	1 year	F	0	45	0	Eggshell index (2.2) - control	Lundholm 1984
Indian runner ducks	1 year	F	4	45	180	Eggshell index (1.6) - S	Lundholm 1984
Indian runner ducks	1 year	F	0	45	0	Calcium secretion (39.4 µg/duck) - control	Lundholm 1984
Indian runner ducks	1 year	F	4	45	180	Calcium secretion (28.0 µg/duck) - S	Lundholm 1984
American kestrels	adult	F	0	168	0	Eggshell thickness (0.171 mm) - control	Lincer 1975
American kestrels	adult	F	0.05	168	8	Eggshell thickness (0.175 mm) - NS	Lincer 1975
American kestrels	adult	F	0.5	168	84	Eggshell thickness (0.145 mm) - S	Lincer 1975
American kestrels	adult	F	1	168	168	Eggshell thickness (0.135 mm) - S	Lincer 1975
American kestrels	adult	F	1.7	168	286	Eggshell thickness (0.126 mm) - S	Lincer 1975

chronic DDE exposure in birds. The near threshold dietary intake of 0.5 mg/kg-day in a sensitive carnivorous species is a more appropriate basis for a maximum tolerable daily intake (TDI) than the square root of the product of the shell thinning LOEL in ducks and an estimated (5.6 times less) shell thinning NOEL. The TDI is more appropriately based on 0.5 mg/kg-day and not on the 0.13 mg/kg-day value used by Environment Canada.

Environment Canada used an uncertainty factor of 10 to account for interspecies variability. A factor of 10 from ducks to sensitive fish-eating raptors is certainly less protective than a factor of 10 from sparrow hawks to sensitive fish-eating raptors. The Lincer (1975) study evaluated the most sensitive chronic endpoint, gender and life stage in a sensitive species. For example, Newton and Bogan (1978) in their report on the DDE-eggshell thinning dose-response, stated: "The regression of shell index on log DDE content in the sparrow hawk was similar to those found by other workers for *Falco peregrinus*, *F. mexicanus* and *Pelecanus occidentalis*." In Chapter 3 of this report, the dietary threshold for DDE reproductive effects in osprey was estimated to be 0.3 ppm in fish. This level would correspond to exactly one-tenth of the 0.5 mg/kg-day threshold in the sparrow hawk, which is calculated from a dietary level of 3 ppm. If one accepts the tenfold uncertainty factor for variability in species susceptibility, the one remaining variable to consider is the rate of dietary intake.

Environment Canada applied an additional uncertainty factor to the TDI to account for the species with the maximum food intake per day. They chose Wilson's storm petrel, with a food intake of 0.94 kg food/kg body weight per day. The choice of the species with the highest rate of food intake should be limited to species as sensitive or nearly as sensitive as the most sensitive species. The choice of Wilson's storm petrel is inappropriate, because petrels have not been shown to be anywhere near as sensitive as the osprey, brown pelican, peregrine falcon, or other sensitive species. In addition, Wilson's storm petrel eats fish only as a minor part of its diet. Most of the petrel's diet is at lower trophic levels, explaining, at least in part, the lower sensitivity of this species to the reproductive effects of DDE.

For example, Coulter and Risebrough (1973) measured 43 ppm DDE in ashy petrel eggs that were thinned only 8–9%. The authors concluded: "The magnitude of shell-thinning is apparently less than a critical level that would affect reproductive success." Henny et al. (1982b) measured DDE residues in eggs from Leach's storm petrel collected in 1979 along the Oregon coast. DDE residue levels averaged 2.5 ppm. Eggshell thinning in Leach's storm petrel measured in eggs collected from 1946 to 1979 did not exceed 8%. Pearce et al. (1979) reported residues of DDE in Leach's storm petrel eggs of 0.75–6.81 ppm. The eggs were collected in 1972 and 1976 off the east coast of Canada. The authors reported measuring shell thickness, but no data were published. The authors claimed that 12 ppm DDE in eggs produced 20% shell thinning. This conclusion was based on an extrapolation of the residue shell-thinning data. Again, no data or regression plots were present in the article. Elliot et al. (1989) reported DDE residues in Leach's storm petrel eggs collected off the Pacific coast of Canada in 1970–1985. Residue levels ranged from 0.601 to 2.16 ppm. Residues in eggs of fork-tailed storm petrel eggs ranged from 1.68 to 2.62 ppm. The authors cite the 12 ppm DDE critical level reported by Pearce et al. (1979).

Elliot et al. (1989) concluded that DDE levels were well below concentrations known to reduce reproductive rates or survival in related species elsewhere.

With critical egg residue levels for hatching failure in the range of 3–4 ppm for sensitive species, Wilson's storm petrel appears to be an inappropriate choice for a protective rate of food intake. The protocol document lists many species that are consumers of aquatic biota. In this list, the osprey appears to be the most sensitive species. The daily food intake rate for the osprey is listed as 0.2 kg/kg body weight-day. If one considers both the rate of food intake and reproductive effect threshold to DDE as a measure of sensitivity to DDE, the osprey appears to be the most sensitive among the listed species.

Using the Environment Canada method, the reference concentration can be calculated directly from the ppm DDE in the sparrow hawk diet. If one divides the 3 ppm dietary level, a level that produced 15% shell thinning, by an uncertainty factor of 10, the maximum NOEL for reproduction in the most sensitive species is 0.3 ppm or 300 ppb in the diet. Assuming the osprey is the most sensitive species, with a food consumption rate of 0.2 kg/kg (Table 1 in the Protocol document) and the sparrow hawk with a food consumption rate of 0.167 kg/kg (calculated from data in Appendix 16 of the Assessment document), the reference concentration in fish is 300 ppb × 0.167/0.2 = 250 ppb.

The reference concentration (which becomes the tissue reference guideline or TRG) calculated above is 18 times higher than that recommended by Environment Canada. Environment Canada's 18-fold lower TRG is due to the use of shell thinning instead of hatching failure as the toxic endpoint and using data from an insensitive species for estimating the maximum food intake rate.

The Canadian guidance is 11 times more protective than the 150 ppb fish residue level that is the basis for the National criterion as well as a level protective of osprey derived in this review. Not every community can afford a level of protection as high as that provided by the Canadian guidance. Lower levels of protection have been shown to provide healthy wildlife populations at far less expenditure of limited resources. Continuing declines in fish residues of the DDTs will continue to increase the degree of protection for wildlife.

3.2.5 US EPA Region IX BTAG Fish Guidance to Protect Wildlife

The Navy/US EPA Region 9 Biological Technical Assistance Group (BTAG) developed toxicity reference values (TRVs) for ecological risk assessments. BTAG TRVs have been published by the California Department of Toxic Substances Control (2008). For the DDTs, the low TRV is 0.09 mg/kg-day. This value is borrowed from the US EPA Great Lakes Criteria (US EPA 1995). The US EPA derived the TRV from the Anderson et al. (1975) finding of a LOEL at 150 ppb DDTs in northern anchovy. The TRV is three times more protective than the US EPA national water criterion that also relies on the Anderson et al. (1975) study. The difference in the two US EPA criteria is the use of an uncertainty factor of 3 in the Great Lakes criterion to estimate a NOEL from a LOEL. This value for uncertainty was used despite the authors acknowledging that DDTs in brown pelicans and northern anchovies were likely not at a steady state in the 1975 data from Anderson et al. The Great Lakes study estimated a sevenfold lower egg residue than found by Anderson et al. (1975), if brown pelicans took 2 years to reach steady state with the DDTs in northern anchovy. Two years is a reasonable estimate based on the very slow excretion of the DDTs and the 10–15 year lifespan of the brown pelican, compared to the 1–2 year lifespan of northern anchovy. Consideration of these facts supports the 150 ppb level in fish as a NOEL, when a steady-state is reached. The use of the BTAG fish guidance provides an extra threefold level of protection beyond that provided by the National criterion.

3.2.6 California EPA Sport Fish Guidance for DDT to Protect Human Health

The SARWQCB (2006) has concurred with U.S. EPA Region IX (2002) to use 100 ppb total DDT in fish fillets as a TMDL target to protect human health. The 100 ppb target was adopted from guidance issued by the Office of Environmental Health Hazard Assessment (OEHHA) of the California EPA. The guidance was developed to protect sport fishermen. The guidance is explained in a report published by OEHHA scientists in 1991 (Pollock et al. 1991) and a later update (Brodberg and Pollock 1999). The following is a review of the sport fish guidance developed by OEHHA.

The guidance was based on fish caught in Southern California in 1987. The focus was the high concentrations of total DDT in fish in the area of the Palos Verdes Shelf. Fish there were contaminated from DDT wastes from the Montrose Chemical Company that were released by way of the Los Angeles County Sanitation District's wastewater outfall at White's Point. The intent was to limit the potential cancer risks of ingestion of a variety of fish species at the more highly contaminated sites.

A trigger level, set at a lifetime cancer risk of 1/100,000, was developed for each chemical based on cancer potency in rodents and assuming a linear dose-response. The OEHHA report stated that:

> The trigger levels for total DDTs and chlordanes are based on excess cancer risks of about 1 in 100,000 (1 x 10^{-5}).

and that:

> Recommendations are provided for species and sites which exceeded 100 ppb of either total DDTs or PCBs or 23 ppb of total chlordane.

The trigger levels were not intended to be used as standards as stated in the report as follows.

> The trigger levels were developed specific to this study, therefore, and should not be used in deriving standards.

Although the trigger levels were developed for each species and chemical, the overall objective was to achieve a potential cancer risk of less than 1/10,000 as stated in the report as follows:

> The specific recommendations for each site and species attempt to reduce exposures to levels that result in overall risks of less than 1 x 10^{-4} (risk for PCBs at the MDL) or lower depending on the site.

This latter objective was overlooked by both US EPA and the SARWQCB in deciding to use the 100 ppb guidance as a TMDL target for total DDT. OEHHA's objective was to have the total cancer risk for a site, considering multiple species and chemicals, below a potential lifetime cancer risk of 1/10,000, not necessarily below a risk of 1/100,000. The 1/100,000 objective was an operational goal by species and chemical and was clearly not intended for adoption as a TMDL target. Considering the levels of chlordane, PCBs and total DDT in fish fillets from Newport Bay (Allen et al. 2004), estimates of potential cancer risks are below 1/10,000, meeting the site objective in the OEHHA guidance. In fact, OEHHA has not issued a fish consumption advisory for Newport Bay.

More recently, OEHHA has revised the fish advisory for DDT (Klasing and Brodberg 2008). The recommendation is an tissue advisory level of 520 ppb for DDT. This guidance weighs the cancer and noncancer risks of DDT against the benefits of eating fish.

The risk of cancer from exposure to DDTs is inappropriately estimated by extrapolation of rodent tumor dose-response with the linearized multi-stage model. This model is intended for use with genotoxic carcinogens. The weight of evidence indicates that DDTs are not genotoxic. This point is made for DDE in a chapter on carcinogenesis (Pitot and Dragan 1996) in the most widely used text in toxicology. The authors indicated that DDE is not mutagenic and acts as a promoter. This conclusion is further explained in a publication from the Pitot laboratory (Holsapple et al. 2006) as quoted below.

> **Mode of action and human relevance of phenobarbital-like rodent liver carcinogens.**
> Phenobarbital is the prototype of several rodent hapatocarcinogens (e.g., oxazepam, DDT) that induce tumors by a non-genotoxic mechanism involving liver hyperplasia (Williams and Whysner, 1996).

The threshold for promotion is orders of magnitude higher than that for a significant carcinogenic risk estimated by the linearized multistage model. Hence, the linear extrapolation risk numbers in the OEHHA guidance overestimate the actual cancer risk. The potential for overestimating the cancer risks is acknowledged in the OEHHA guidance.

V.A.5.a.(1). DDTs, Chlordane, and PCBs. The classification of DDTs, chlordane, and PCBs as potential (probable) human carcinogens is based on animal studies conducted using high doses of the chemicals. Some scientists may argue that DDTs and PCBs are not tumor initiators but rather, promoters. Resolution of this debate is beyond the scope of this report. We also recognize that the derivation of the carcinogenic potency factors (CPF or Q_1^*) are based on numerous assumptions.

Overall, the assumptions used to derive the CPF are weighted such
that the estimated cancer risk at a given dose is unlikely to be
higher than estimated, but most likely will be lower (maybe by orders
of magnitude) and perhaps may even be zero.

These concepts were known as early as the late 1960s, explaining, in part, why the U. S. Food and Drug Administration set the action level for DDTs in commercial fish at 5,000 ppb. That action level is still in effect today (US FDA 2007).

The OEHHA guidance dealing with the risk of human cancer from ingestion of sport fish fillets has been misinterpreted to claim impairment of beneficial uses of Newport Bay. However, even the 1/100,000 potential risk level is met by those ingesting sport fish from Newport Bay. As reported in the Allen et al. (2004) study, a survey among local anglers identified the most sought after species of fish. Four of the top five were analyzed for DDTs. Total DDT residues in these four species by preference rank were 69, 68, 64 and 84 ppb. The average DDT residue in 14 species of sport fish was 79 ppb. These fish were captured in 2000 and 2001. The levels today are almost certainly lower. Considering these residue levels in sport fish fillets, the 520 ppb target is met and even the older guidance of 100 ppb is met. There is no impairment of sport fishing in Newport Bay.

4 Chlordane

4.1 Levels in the Environment

The following sections present available chlordane data for Newport Bay and Watershed. Downward trends in chlordane concentrations—particularly in fish tissue and mussel tissue—are evident in data collected for almost 20 years.

4.1.1 Agricultural Soils

The half-life of chlordane in soil is estimated at 350 days (or approximately 1 year), but can range from 37 to 3,500 days (approximately 10 years) (Hornsby et al. 1996). Chlordane is persistent in soils and volatilization is believed to be the major removal mechanism (US DHHS 1994). Chlordane data for agricultural soil are available for the Newport Bay watershed for the years 1989, 1990, 1995, 2000, 2002, 2004 and 2006 (Table 28).

Samples were taken from different locations and different years with the purpose of assessing site conditions for planning and development and not to establish concentration trends over time in the Watershed. The vast majority (approximately 95%) of chlordane soil samples returned concentrations below detection limits. Detectable concentrations ranged between 47 and 240 ppb.

Table 28 Chlordane concentrations in agricultural soils in the Newport Bay Watershed

Year	0-12 inch Sample Depth			12-24 inch Sample Depth			>24 inch Sample Depth			Detection Limits (ppm)
	Range of Detected Chlordane (ppm)	Total Samples	Total Non-detect Samples	Range of Detected Chlordane (ppm)	Total Samples	Total Non-detect Samples	Range of Detected Chlordane (ppm)	Total Samples	Total Non-detect Samples	
1989	0.240 - 0.240	3	2	--	1	1	0.120 - 0.190	5	3	0.08 or 0.12
1990	0.170 - 0.170	2	1	0.210 - 0.210	1	0	0.190 - 0.190	3	2	0.08 or 0.12
1995	0.047 - 0.055	24	22							0.03
2000	--	28	28							0.05 or 0.50
2002	0.050 - 0.130	174	161				--	27	27	--
2004	--	230	230				--	45	45	0.1 - 10.0
2006	0.077 - 0.160	6	1				--	6	6	0.05

Sources: Unpublished technical reports provided by The Irvine Company (1985–2006)
Data were not available for shaded areas

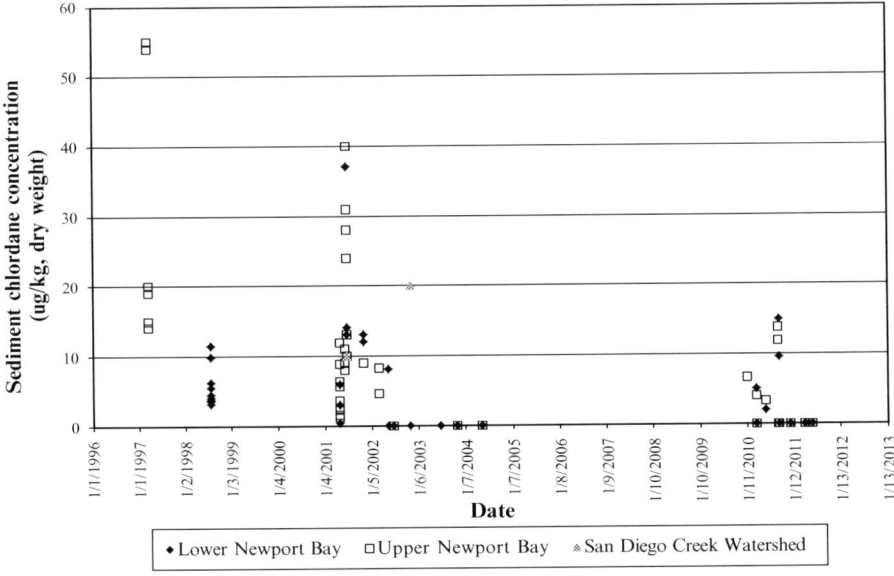

Fig. 25 Chlordane concentrations above method detection limits in sediments from Newport Bay and Watershed (1997–2011). *Sources*: Masters and Inman (2000); SCCWRP BIGHT 98 Survey (1998); Bay and Greenstein (2003); Bay et al. (2004); unpublished technical reports provided by The Irvine Company (2000–2004); SARWQCB (2006); Orange County Watersheds (2010–2011)

4.1.2 Sediments

Chlordane levels in Bay and Creek sediments are available for lower Newport Bay, upper Newport Bay, and San Diego Creek for the period 1997 through 2011 (Fig. 25). As with DDT, it is difficult to infer trends in sediment chlordane concentration over time from these data for several reasons. First, sampling was conducted by multiple agencies, using multiple methodologies, at varying locations and sample depths. Given this diversity in sampling approach and location, direct comparisons among data from year to year are inappropriate. Second, there is significant movement of sediment into, out of, and within the Bay and its Watershed such that even samples taken in the same location at two different times may not represent the change in chlordane concentration for a specific quantity of sediment. Sediment movement resulted both from the natural flow of water and sediment in the Bay and its Watershed, as well as from periodic dredging in the Bay, which occurred in 1983, 1985, 1988, and 1999. Third, sediment concentrations in Newport Bay may be more indicative of chlordane loads from years or decades past, since Bay sediments are transported from the upper watershed in a highly variable, episodic manner, correlated with storm events and wetter-than-average rainfall years. Thus, chlordane concentrations in Bay sediments reflect chlordane that was applied many years ago in the upper watershed, and then sorbed to sediments in that location, which were subsequently eroded into a creek channel and transported to the Bay. For all these reasons, the available sediment data for Newport Bay are not the most reliable indicators of bioavailable chlordane concentration trends in the watershed. However, it is notable that since 2002 Bay and Creek sediment samples have exhibited chlordane concentrations below 20 ppb.

4.1.3 Water Column

Data from 1998 to 2009 reveal a range of chlordane concentrations in water from Newport Bay and Watershed (Bay and Greenstein 2003; Bay et al. 2004, Orange County Watersheds).[7] Ten of 91 samples were above detection limits with a range from 2.5 to 186.5 pptr (ng/L). Detection limits ranged from 1 to 130 pptr. Detection limits were not reported for 8/10 of the samples above the detection limit, including the highest five values (22.5–130 pptr). The accuracy of chlordane analysis in water at the pptr level is questionable due to chromatographic interference caused by high background levels of organic matter and the binding of chlordane to microscopic (filterable) macromolecular organic matter.

4.1.4 Fish and Mussels

Red shiner tissue concentrations may be taken as an indicator of chlordane concentrations in the Watershed, as red shiners are local, short-lived species. For this species, chlordane tissue concentrations dating from 1983 show a substantial decline over time. First-order decay constants were derived using historical red shiner tissue data for the entire time period (1983–2000) and for two subsets of the data set (1983–1991 and 1992–2000). The equations of these decay curves are indicated in Figs. 26 and 27. Figure 27 also shows projected chlordane concentrations through 2010. Data are not available to confirm these concentrations.

When all red shiner data are considered together, a statistically strong (R^2 value of 0.774) downward trend in chlordane concentration is evident. Exponential decay curves fit to the two subsets of data revealed consistent downward trends during

---

[7] Personal communication from Amanda Carr at Orange County Watersheds.

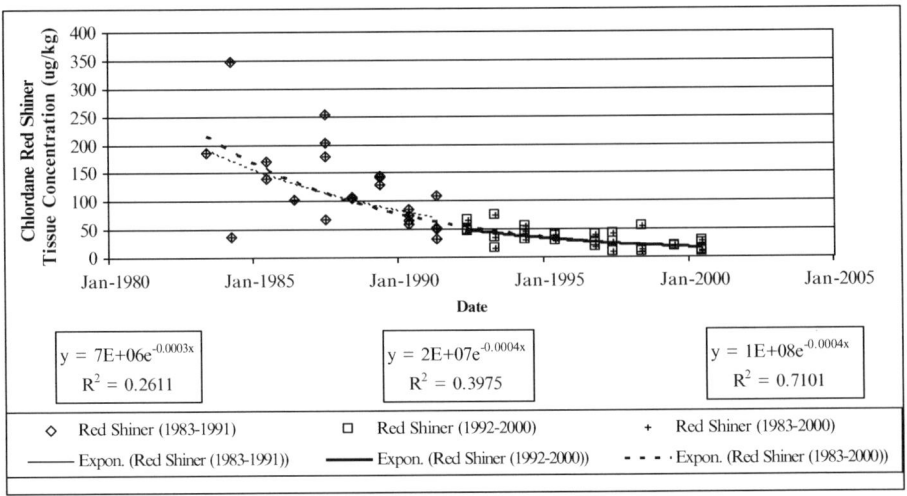

Fig. 26 Chlordane concentrations in red shiner, San Diego Creek and Peters Canyon Wash (1983–2000). Data from California Toxic Substances Monitoring Program (1983–2000). Red shiner data from 2002 to the present are not available

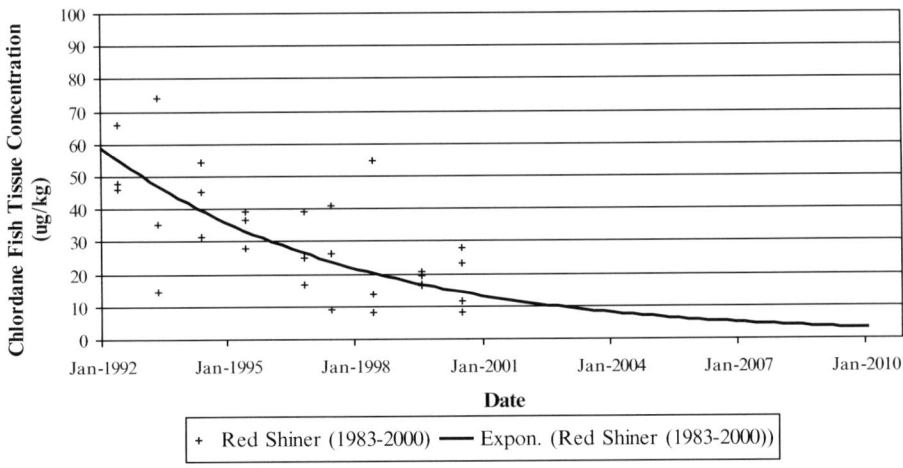

Fig. 27 Chlordane concentrations in red shiner in San Diego Creek and Peters Canyon Wash projected through 2010. Data from California Toxic Substances Monitoring Program (1983–2000). Red shiner data from 2002 to the present are not available

both periods. Therefore, the downward trend observed in the complete data set (1983–2000) is not simply the result of a temporally localized effect, but rather is an accurate portrayal of declines in chlordane concentrations over the entire period. The decay rate (−0.00046 per day or −0.17 per year) obtained for the full red shiner data set (1983–2000) is equivalent to a half-life of 4.1 years.

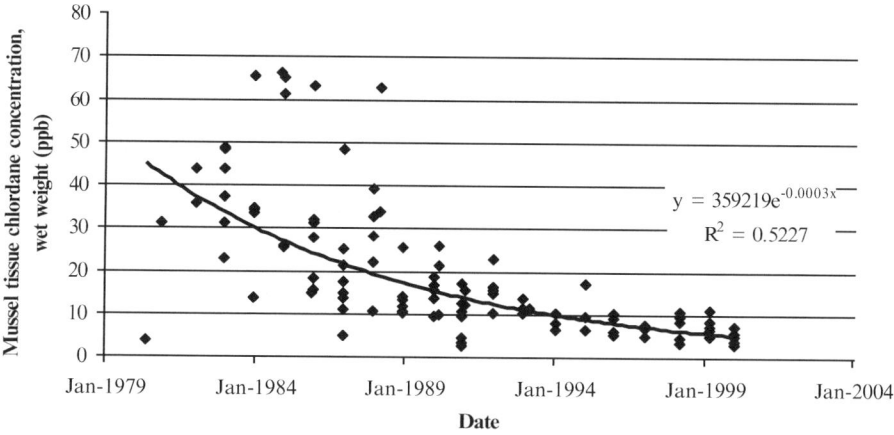

Fig. 28 Chlordane concentrations in mussels from Newport Bay and Watershed. Data from California Mussel Watch Program, 1980–2000. The Mussel Watch Program was ended in 2000

Like red shiner data, mussel tissue data dating to 1980 from Newport Bay show decreasing chlordane concentrations (Fig. 28). An exponential regression analysis of mussel data (by wet weight) for the period of record (1980–2000) showed a reasonably strong chlordane concentration decline rate in mussels ($R^2 = 0.5227$). A split analysis was also performed on mussel data for the two periods 1980–1989 and 1990–2000. Although the split analysis indicated that neither the earlier ($R^2 = 0.1176$), nor the later period ($R^2 = 0.2968$) demonstrate as statistically strong a decline as the complete period, the entire mussel data set (1980–2000) reflects a statistically significant decline in chlordane tissue concentrations that is equivalent to a half-life for chlordane of 6.2 years (decay rate of −0.00031 per day or −0.11 per year).

4.2 Benthic Triad Analysis of Impairment

The requirement for a TMDL for chlordane in Newport Bay is based on a triad assessment of sediment chemistry, sediment toxicity to benthic organisms and degradation of benthic communities. In this section, we examine the science underlying the triad to determine whether the results indicate impairment to benthic organisms in Newport Bay. The starting point is a discussion of the assays that make up the triad.

The sediment triad has three components. The first component is the concentration of chlordane in sediments from Newport Bay. The SARWQCB staff report contains data on chlordane levels in sediments from several reports (Appendix A-2 to SARWQCB 2006). Levels of chlordane in sediments from Newport Bay ranged from <1 to 55 ppb in the time period 1994–2011 (see Fig. 25 above). The overall average appears to be less than 10 ppb. As noted above, chlordane levels are declining in the Watershed.

Table 29 Correlation of sediment concentrations of metals and organics with amphipod toxicity. Table 30 reproduced from the Bay Protection and Toxic Cleanup Program report (CSWRCB 1998) Table 30 Spearman Rank Correlation results for selected toxicants significantly correlated with amphipod toxicity (Eohaustorius and Rhepoxynius) results from specific water bodies

Water Body	Chemical	N	Spearman Rho	Significance
Anaheim Bay	Selenium	22	−0.453	0.025
Huntington Harbor	Antimony	15	−0.757	0.001
Huntington Harbor	Lead	15	−0.629	0.01
Huntington Harbor	Tin	15	−0.842	0.0005
Newport Bay	Percent Fines	20	−0.649	0.0025
Newport Bay	TOC	20	−0.422	0.05
Newport Bay	Antimony	20	−0.458	0.025
Newport Bay	Chromium	20	−0.598	0.005
Newport Bay	Copper	20	−0.542	0.01
Newport Bay	Lead	20	−0.392	0.05
Newport Bay	Mercury	20	−0.444	0.05
Newport Bay	Nickel	20	−0.633	0.0025
Newport Bay	Tin	20	−0.495	0.025
Newport Bay	Zinc	20	−0.497	0.025
Newport Bay	Total Chlordane	20	−0.380	0.05
Newport Bay	Total PCB	20	−0.408	0.05

4.2.1 Correlations of Sediment Residues and Benthic Toxicity

The second component in the triad is toxicity of sediments to benthic organisms. Two bioassays, used extensively in Newport Bay studies, are mortality to amphipods and sediment pore water inhibition of fertilization and larval development in purple sea urchins. Chlordane was negatively correlated with amphipod survival for 20 sampling sites in Newport Bay as shown in Table 29, which is reproduced above from the Bay Protection and Toxic Cleanup Program report (CSWRCB 1998).

The correlation coefficient was −0.38, and was significant at the 0.05 level. Higher correlations were observed in the same samples with percent fines, total organic carbon, and eight metals. The chlordane residue level was not correlated with inhibition of purple sea urchin fertilization or larval development in sediment pore water samples from Newport Bay (SWRCB 1998). The chlordane residue level was correlated with inhibition of purple sea urchin larval development in data collected in the entirety of the area regulated by the SARWQCB (Region 8), but the report did not indicate whether this correlation applied to Newport Bay. Many of the sediment samples from Newport Bay contained levels of ammonia and sulfide that were toxic in the amphipod and purple sea urchin bioassays. Hence, the toxicity observed in many of the sediments was due to ammonia and sulfide.

The third component of the triad is benthic community degradation. Crustaceans are generally the most sensitive species in the benthos and are given extra weight in the benthic index. The benthic index was not reported to be correlated with chlordane concentrations in sediments from Newport Bay (Table 29 in CSWRCB 1998).

Overall, chlordane concentrations in sediments from Newport Bay are weakly correlated with toxicity to organisms in the benthos. This correlation is not by itself an indication of causation. The toxicity correlated with chlordane levels could well be explained by the presence of metals that also correlated with toxicity or by the hundreds of other chemicals that could be present, but were not measured. Other investigators (Bay et al. 2004) have suggested that amphipod toxicity is associated with the existence in the sediment and water of unmeasured organic compounds, possibly organophosphorus or pyrethroid insecticides. The authors of the Bay Protection and Toxic Cleanup Program report (CSWRCB 1998) also noted the presence of uncharacterized organic compounds that could have contributed to sediment toxicity.

4.2.2 Effects Range Median (ERM)

Since the SARWQCB staff report (2006) compared chlordane levels to the ERM (Long and Morgan 1990), the data underlying the ERM for chlordane becomes important to any conclusions reached by the triad analysis of impairment. The ERM is calculated from 12 data points reproduced below (Table 30) from Long and Morgan (1990).

The first two data points (0.3 and 0.6 ppb chlordane) were derived by equilibrium partitioning from the chronic marine CTR standard for water to sediment using the lower 95th and 99th percentile of the variability in K_{oc} values (Pavlou et al. 1987). These two data points are in error for two reasons. The first reason is that the K_{oc}

Table 30 Twelve data points used to calculate the ERM for chlordane. Table 33 reproduced from Long and Morgan (1990) Table 33 Effects range-low and effects range-median values for chlordane and 12 concentrations used to determine these values arranged in ascending order

Concentrations (ppb)	End Point
0.3	EP 99 percentile chronic marine
0.5	ER-L
0.6	EP 95 percentile chronic marine
2.0	San Francisco Bay, California, AET
3.5	San Francisco Bay, California, bioassay COA
3.5	San Francisco Bay, California, bioassay COA
4.1	San Francisco Bay, California, bioassay COA
6.0	ER-M
6.4	San Francisco Bay, California, bioassay COA
17.4	EP freshwater lethal threshold
25.0	DuPage River, Illinois, benthos COA
31.3	Trinity River, Texas, bioassay COA
120.0	SSB LC50 for *C. septemspinosa*
<5,800.0	SSB LC50 for *N. virens*

values used are outdated, because they are not based on the superior slow-stir technique (de Bruijn et al. 1989). Second, the high variability in the outdated K_{oc} values is not seen in those derived by the slow-stir method, precluding the necessity of using the lower 95th and 99th percentile of the K_{oc} values. Multiplying the K_{oc} value for chlordane published by US EPA Region IX (2002) and the SARWQCB staff (SARWQCB 2006) by the chronic marine CTR standard gives a single data point of 65 ppb chlordane in sediment containing 1% OC.

The last two data points in Table 30 are based on spiked sediment bioassays (McLeese and Metcalfe 1980; McLeese et al. 1982). These two bioassays are used to assess toxicity primarily from the water column rather than from sediment. The first study was done with sand shrimp (*Crangon septemspinosa*) and involved adding an unreported amount of chlordane to a beaker, drying off the solvent, adding water and coarse sand (0.28% OC; 0.5–2 mm diameter particles). The sand was allowed to settle and the shrimp were added. The authors concluded that chlordane dissolved in the water phase was the primary cause of toxicity. Chlordane bound to sediments contributed little to toxicity. For these reasons, and the fact that the chlordane moved from water to sediment, this bioassay is primarily a water bioassay. The same is true of the second study (McLeese et al. 1982). The difference between the two studies is that in the second study the organism was a polychaete worm (*Nereis virens*) and the sediment was sandy silt that contained 2% OC. In a true sediment bioassay, all of the chlordane would be picked up off of the glass by the sediment. The sediment would then be transferred to a clean container and equilibrated with water. Samples of water and sediment would be analyzed periodically until an equilibrium was reached. The test organism would be added only after equilibrium was achieved.

Sediment LC_{50}s for these two data points can be estimated using equilibrium partitioning. If one applies the K_{oc} published in the U.S. EPA Region IX (2002) and SARWQCB staff reports (2006) and the water only LC_{50}s reported by McLeese et al. (1982), the estimated sediment LC_{50}s are 9,000 ppb for sand shrimp and 7,100,000 ppb for the polychaete worm.

The remaining 8 data points in Table 33 from Long and Morgan (1990) are based on the presence of chlordane (along with potentially hundreds of other chemicals) in toxic sediments. None of these eight data points provide dose-response information.

The flaws in the ERM data set preclude the use of the ERM value as an indication of the threshold for benthic toxicity due to chlordane in sediments. The threshold appears to be orders of magnitude greater than the ERM. This conclusion is further supported by other bioassay data.

4.2.3 Equilibrium Partition Estimates of Toxicity Thresholds

Let us look further at what is known about toxicity thresholds for chlordane to amphipods and other benthic organisms to gain an understanding of whether the levels of chlordane in sediments in Newport Bay are high enough to cause toxicity to these organisms. Cardwell et al. (1977) studied the chronic toxicity of chlordane in the amphipod, *Hyallela azteca* (Table 31).

Table 31 Growth and survival of *Hyallela azteca* exposed in the water column to chlordane. Table 23 reproduced from Cardwell et al. (1977)
Table 23 Relative survival and growth of Hyallela azteca exposed to technical chlordane

Parameter	Measured concentration of technical chlordane, µg/L					
	Control	1.4	2.6	5.3	11.5	20.5
Replicate I						
No. survivors[a]	27	23	23	24	3	0
% Survivors	108	92	92	96	12	0
Wet body weight, mg[b]	6.3±1.3	6.2±1.5	6.4±1.2	5.1±0.9	3.8±0.7	...
Dry weight, mg[b]	1.58	1.49	1.57	1.37	0.87	...
Replicate II						
No. survivors[a]	22	25	25	24	9	0
% Survivors	88	100	100	96	36	0
Wet body weight, mg[b]	7.5±1.3	5.8±1.3	5.8±1.6	5.5±1.6	5.3±1.0	...
Dry weight, mg[b]	1.92	1.55	1.53	1.35	1.33	...

[a]25 individuals introduced initially per chamber
[b]Average calculated weight per individual

In a 65 day study of mortality and weight gain, the NOEL appears to be 2.6 ppb in water. Using equilibrium partitioning to estimate the sediment concentration of chlordane required to reach 2.6 ppb in water, gives a sediment level of 42,172 ppb at 1% OC (2.6 µg/L × 1,622,000 L/kg × 0.01 µg OC/µg sediment = 42,172 µg/kg).

The US Fish and Wildlife Service (Eisler 1990) reviewed the aquatic toxicity of chlordane. Eisler reported an LC_{50} of 40 ppb for the amphipod *Gammarus fasciatus*. The equivalent LC_{50} for sediment at equilibrium would be 649,000 ppb. Other sensitive aquatic species include the pink shrimp (LC_{10} of 0.24 ppb in water), planarian (5 day NOEL of 0.2 ppb in water), and dungeness crab survival and molting (37 day NOEL of 0.015 ppb in water). The dungeness crab bioassay appears to be the most sensitive; the equilibrium NOEL in sediment calculated out to be 243 ppb chlordane in sediment.

The triad analysis, although representing one kind of weight-of-evidence analysis, is incomplete and flawed as it was used to assess impairment of aquatic biota by chlordane in Newport Bay. Relying on the mere presence of chlordane along with hundreds of other chemicals in toxic sediments constitutes an incomplete weight-of-evidence analysis. The chlordane ERM is not a reliable measure of toxicity thresholds and should not be used in a weight-of-evidence analysis to assess impairment of aquatic biota. Most importantly, one should consider the results of dose-response bioassays. Valid spiked sediment bioassays could not be found for chlordane. Therefore, the triad analysis should have relied on spiked water bioassays and equilibrium partitioning to estimate toxicity thresholds for chlordane in sediments. The available aquatic toxicity bioassay data do not support an effect of chlordane on benthic organisms at the level of approximately 10 ppb (<1–55 ppb) as measured in sediments from Newport Bay. The lowest effect level exceeded 1,000 ppb and the NOEL in the most sensitive species and life stage was 243 ppb.

5 Toxaphene

5.1 Levels in the Environment

The following sections present available toxaphene data for Newport Bay and Watershed. Trends in toxaphene concentrations—particularly fish tissue concentrations—are evident in data collected for 20 years in Newport Bay and Watershed.

5.1.1 Agricultural Soils

For toxaphene there were fewer agricultural soil data available than for DDT. As with DDT data, samples from different years were taken in different locations since the purpose of sampling was to assess site conditions for planning and development purposes, not to establish concentration trends over time in the watershed. The majority of toxaphene soil samples returned concentrations below detection limits (Table 32). For example, for 2004 data, all 275 soil samples yielded concentrations below the analytical detection limit of 0.1 ppm. Although no statistically clear trends in soil toxaphene concentrations can be demonstrated from a data set in which roughly 90% of samples have toxaphene concentrations below detection limits, and in which samples were not taken at the same locations over time, it appears from these data that the mass of toxaphene in the watershed is currently quite small. This is consistent with expectations based upon the half-life of toxaphene, as detailed below.

5.1.2 Sediments

Toxaphene was detected in only one of 42 sediment samples collected from 2005 to 2008.[8] The one sample had a toxaphene residue of 31.7 ppb. The detection limit for all 42 samples was 10 ppb. The most recent analysis of sediments in 2011 (Orange County Watersheds 2013) found no toxaphene above high detection limits that ranged from 70 to 1,900 ppb.

Table 32 Toxaphene concentrations in agricultural soils in the Newport Bay Watershed

Year	0-12 inch Sample Depth			12-24 inch Sample Depth			>24 inch Sample Depth			Detection Limits (ppm)
	Range of Detected Toxaphene (ppm)	Total Samples	Total Non-detect Samples	Range of Detected Toxaphene (ppm)	Total Samples	Total Non-detect Samples	Range of Detected Toxaphene (ppm)	Total Samples	Total Non-detect Samples	
1989	0.500 - 0.940	3	0	0.540 - 0.550	3	1	--	3	3	0.25
1990	0.190 - 0.220	2	0	--	1	1	--	3	3	0.16
1995	--	19	19							0.06
2000	--	28	28							0.20 & 2.00
2002	0.340 - 2.300	174	125				0.210 - 0.300	27	24	--
2004	--	230	230				--	45	45	0.10 - 10.0
2006	--	6	6				--	6	6	0.10

Sources: Unpublished technical reports provided by The Irvine Company (1989–2006)
Data were not available for shaded areas

---

[8] Personal communication from Amanda Carr at Orange County Watersheds.

5.1.3 Water Column

Ninety water samples collected from 2002 to 2009 have been analyzed for toxaphene (Bay and Greenstein 2003; Orange County Watersheds (see footnote 8)). Only one sample contained detectable toxaphene. The detected level was 5.5 pptr, which was less than the stated detection limit of 10 pptr.

5.1.4 Fish and Mussels

In the case of red shiner, toxaphene tissue concentration data dating from 1983 show a substantial decline (see Figs. 29 and 30). Red shiner may be taken as an indicator of toxaphene concentrations in receiving waters within the watershed, as red shiners are a local, short-lived species. The primary statistical approach to establishing the declining trend in toxaphene concentrations in the watershed has been to derive first-order decay constants using historical toxaphene data for red shiner fish tissue. The equations of these curves are indicated in Fig. 29.

When all red shiner data are considered together, a statistically strong (R^2 value of 0.671) downward trend in toxaphene concentration is evident. The statistical analysis that characterizes these trends was confirmed by splitting the data set for red shiners into two separate sets consisting of the first 10 years of data (1983–1992) and the second 10 years of data (1993–2001). Calculated first-order decay rates for the red shiner toxaphene data are statistically similar for the full data set and for the sub-sampled datasets. The decay rate (−0.00055 per day or −0.20 per year) obtained

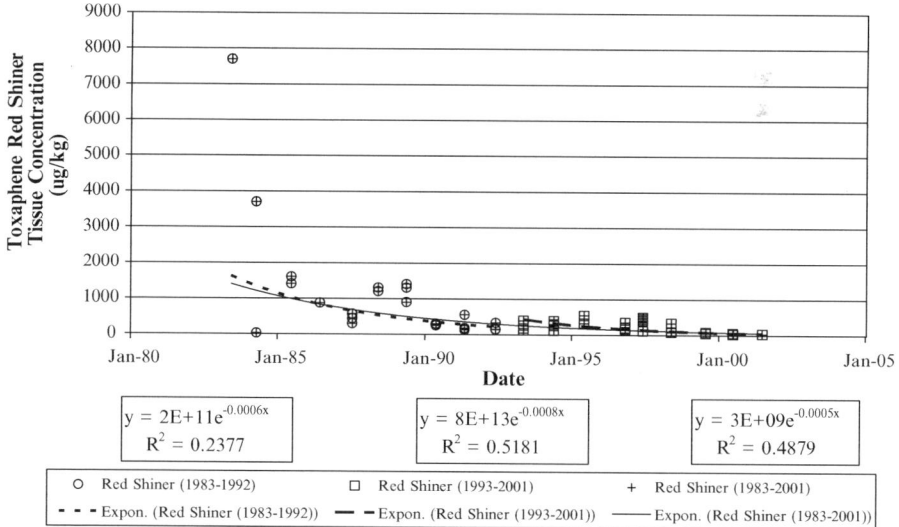

Fig. 29 Toxaphene concentrations in red shiner, San Diego Creek and Peters Canyon Wash (1983–2001). *Source*: California Toxic Substances Monitoring Program (1983–2002). Red shiner data from 2002 to the present are not available

for the full red shiner dataset (1983–2001) is equivalent to a half-life of 3.4 years for toxaphene in the watershed.

Mussel tissue data from Newport Bay for the period 1980 through 2000 do not show any statistically significant trends in wet weight toxaphene concentrations over time (Fig. 31). However, toxaphene concentrations in 84 out of the 111 samples (76%) collected over the 21-year period were below analytical detection limits, and the frequency of non-detect results was consistent over time. Nondetect samples were not plotted in Fig. 31.

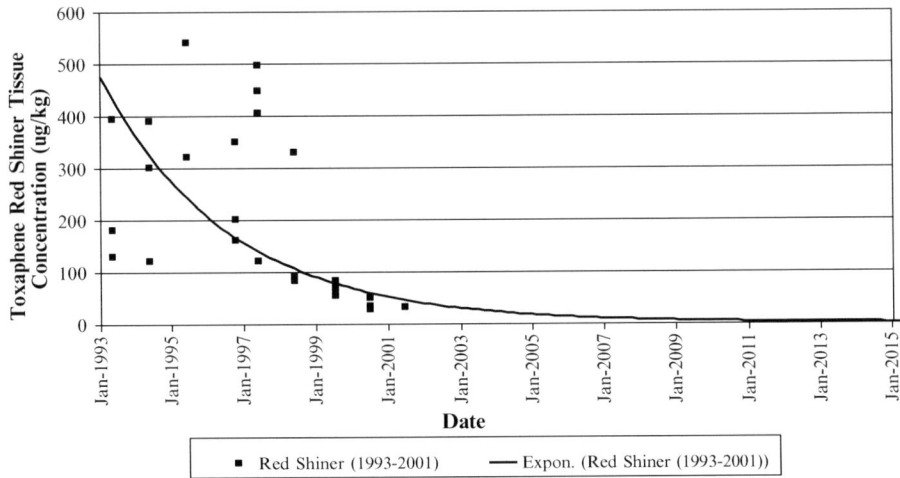

Fig. 30 Toxaphene concentrations in red shiner, San Diego Creek and Peters Canyon Wash, projected through 2015. Data from California Toxic Substances Monitoring Program (1983–2002). Red shiner data from 2002 to the present are not available

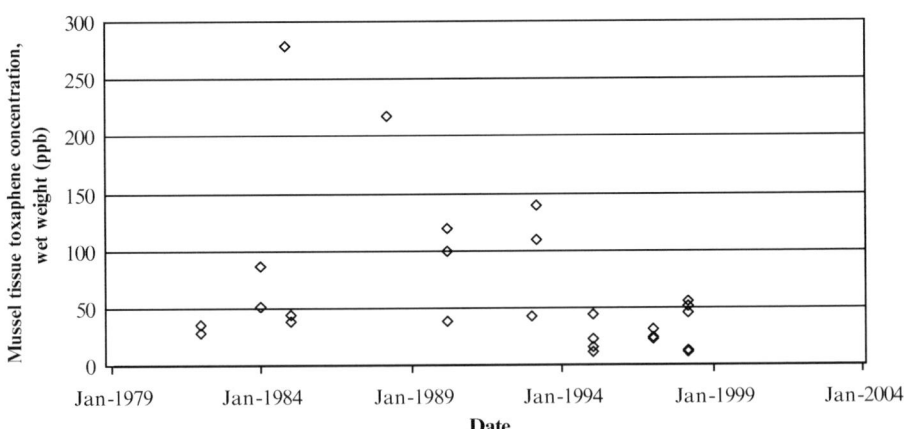

Fig. 31 Toxaphene concentrations in mussels from Newport Bay and Watershed. Data from the California Mussel Watch Program (1980–2000). Only data above detection limits were plotted. The Mussel Watch Program ended in 2000

5.1.5 Factors Affecting Decay of Toxaphene Residues

The U.S. EPA banned the use of Toxaphene in 1990. The observed decline in toxaphene concentrations in fish tissue and the low observed toxaphene concentrations in watershed soils and sediments are partly attributable to the natural removal of toxaphene from the watershed. The half-life of toxaphene in soil is reported as ranging from 1 to 14 years (US EPA 1999). The wide range is attributable to apparently differing degradation rates for toxaphene under aerobic and anaerobic conditions (US DHHS 1996). Under anaerobic conditions the half-life of toxaphene in soil and sediment has been reported to be on the order of weeks to months (Callahan et al. 1979). However, under aerobic soil conditions, Nash and Woolson (1967) reported a half-life of 11 years. Assuming that aerobic conditions are most common in the Newport Bay watershed suggests a half-life on the order of 11 years. At this rate and given that the use of toxaphene was banned in 1990 and excluding other loss mechanisms, the mass of toxaphene in the agricultural soils of Newport Bay and Watershed would have declined by at least 63% over the past 16 years due solely to natural removal. Assuming anaerobic conditions—conditions typical of sediments submerged in water, such as bay sediments—the half-life of toxaphene is on the order of weeks or months. This suggests sediments, most of which remain consistently submerged in the watershed, should currently contain very little toxaphene. The half-life for toxaphene in the watershed that was estimated using red shiner fish tissue data (3.4 years) is consistent with these estimates for the half-life of toxaphene in watershed soils.

5.2 NAS Fish Guidance to Protect Wildlife

The SARWQCB (2006) have decided to use the NAS (1972) guidance of 100 ppb toxaphene in fish to protect wildlife consuming fresh water fish in the Watershed. As is apparent from viewing the regression in Fig. 30, the current level in red shiners is well below this guidance. Even so, the SARWQCB has chosen to base impairment on the older fish data that exceed the guidance and to require a sediment target to achieve the fish target of 100 ppb.

5.3 New York State Sediment Guidance to Protect Wildlife

US EPA Region IX (2002) and SARWQCB (2006) staff have decided to use the New York State Department of Conservation (1998) screening level of 0.1 ppb as a TMDL target for toxaphene in sediments. The following is a detailed look at the scientific basis for this guidance.

New York State chose the equilibrium partitioning method for the derivation of a sediment screening level for toxaphene (New York State Department of Environmental Conservation 1998). The calculation began with a New York State water column criterion of 0.005 ppb (µg/L) toxaphene. Multiplication by a K_{oc} for toxaphene gave the sediment OC concentration. Division by 100 gave the sediment screening level at 1% OC. The K_{oc} was assumed to be equal to a K_{ow} value of 1,995 L/kg (log K_{ow} = 3.3). Hence, the calculation of the screening level is 0.005 µg/L × 1,995 L/kg × 0.01 = 0.1 µg/kg (ppb).

A specific reference for the log K_{ow} of 3.3 could not be found in the New York State Department of Environmental Conservation (1998) guidance document. The K_{ow} used by New York State is 158-fold lower than the K_{ow} cited in the US EPA Region IX (2002) and SARWQCB (2006) reports. Using the EPA and SARWQCB-approved K_{ow} value in the New York State method gives a sediment screening level of 15.8 ppb.

The consequences of using the 0.1 ppb sediment target for toxaphene are major. The SARWQCB (2006) report estimated a Newport Bay Watershed loading capacity of 5.67 g of toxaphene per year and an existing load of 536 g of toxaphene per year. Assuming these loads to be correct, there would have to be a 99% reduction in the toxaphene load to meet the TMDL. Since almost all of the toxaphene is bound to sediment, the sediment load reduction in the watershed would also have to be 99%. A sediment load reduction of this magnitude would be impractical in a major storm event. However, if one uses the 158-fold higher log K_{ow} of 5.5 published by both the US EPA Region IX (2002) and the SARWQCB (2006), the New York State sediment screening level would increase 158-fold. The sediment target would then be 15.8 ppb, resulting in a loading capacity of 896 g per year, a value greater than the estimated existing load of 536 g per year, thereby obviating the need for a TMDL.

6 PCBs

The need for a PCB TMDL was stated by the SARWQCB (2006) in their report on organochlorine TMDLs for Newport Bay and Watershed to be due to the level of PCB residues in sport fish in Newport Bay exceeding guidance levels of 20 ppb recommended by California EPA. This section will take a detailed look at the scientific basis for the impairment analysis. The starting point is a review of the PCB residue data in sport fish.

6.1 Levels in Sport Fish

The SARWQCB organochlorine TMDL report (2006) contains the PCB data (Appendix A-1, Table 1) for sport fish that were relied upon by the SARWQCB in their impairment analysis. The underlying data from the State Board is listed below in Table 33.

Of the 63 fish composites from 16 species of sport fish that were analyzed, 18 had measureable residues of PCBs. Two of these were at or below the reporting limit of 5 ppb, leaving 16 composites above the reporting limit. Nine of the fish

Table 33 Total PCB residues in fish fillets from Newport Bay, 1995–2002

Study	Location	Season	Year	Species	Fish per composite	PCBs (ppb)
CFCP	Lower NB	Summer	1999	Diamond turbot	5	5
CFCP	Lower NB	Summer	1999	Shiner surfperch	5	48
CFCP	Lower NB	Summer	1999	Shiner surfperch	10	94
CFCP	Lower NB	Summer	1999	Spotted turbot	5	11
CFCP	Lower NB	Summer	1999	Yellowfin croaker	5	30
TSMP	Lower NB	Summer	1995	Black croaker	2	≤ 5
TSMP	Upper NB	Summer	2002	Striped mullet	2	172
TSMP	Upper NB	Summer	2002	Spotted sand bass	1	84
TSMP	Upper NB	Summer	1999	Orangemouth corvina	1	21
TSMP	Upper NB	Summer	2000	California halibut	1	18
TSMP	Upper NB	Summer	2001	Round stingray	1	10
TSMP	Upper NB	Summer	1997	Diamond turbot	3	≤ 5
TSMP	Upper NB	Summer	1998	Brown smoothhound shark	1	≤ 5
SCCWRP	Lower NB	Winter	00/01	Barred sand bass	3	≤ 5
SCCWRP	Lower NB	Winter	00/01	Black perch	3	≤ 5
SCCWRP	Lower NB	Winter	00/01	Black perch	3	≤ 5
SCCWRP	Lower NB	Winter	00/01	Black perch	4	≤ 5
SCCWRP	Lower NB	Winter	00/01	California halibut	4	≤ 5
SCCWRP	Lower NB	Winter	00/01	California halibut	4	≤ 5
SCCWRP	Lower NB	Winter	00/01	C-O sole	4	≤ 5
SCCWRP	Lower NB	Winter	00/01	C-O sole	4	≤ 5
SCCWRP	Lower NB	Winter	00/01	Diamond turbot	6	≤ 5
SCCWRP	Lower NB	Winter	00/01	Diamond turbot	6	≤ 5
SCCWRP	Lower NB	Winter	00/01	Fantail sole	5	≤ 5
SCCWRP	Lower NB	Winter	00/01	Spotted sand bass	3	≤ 5
SCCWRP	Lower NB	Winter	00/01	Spotted turbot	4	≤ 5
SCCWRP	Lower NB	Winter	00/01	Spotted turbot	4	≤ 5
SCCWRP	Lower NB	Winter	00/01	Spotted turbot	4	≤ 5
SCCWRP	Lower NB	Summer	2001	Kelp bass	5	≤ 5
SCCWRP	Lower NB	Summer	2001	Yellowfin croaker	4	≤ 5
SCCWRP	Lower NB	Summer	2001	Yellowfin croaker	4	41.2
SCCWRP	Lower NB	Summer	2001	Yellowfin croaker	4	11.0
SCCWRP	Lower NB	Summer	2001	Yellowfin croaker	4	8.1
SCCWRP	Lower NB	Summer	2001	Yellowfin croaker	4	≤ 5
SCCWRP	Lower NB	Summer	2001	Black perch	3	≤ 5
SCCWRP	Lower NB	Summer	2001	Black perch	3	≤ 5
SCCWRP	Lower NB	Summer	2001	Black perch	3	≤ 5
SCCWRP	Lower NB	Summer	2001	Barred sand bass	3	≤ 5
SCCWRP	Lower NB	Summer	2001	Spotted sand bass	3	≤ 5
SCCWRP	Lower NB	Summer	2001	Spotted sand bass	3	10.4
SCCWRP	Lower NB	Summer	2001	Spotted sand bass	3	24.2

(continued)

Table 33 (continued)

Study	Location	Season	Year	Species	Fish per composite	PCBs (ppb)
SCCWRP	Lower NB	Summer	2001	California corbina	3	≤5
SCCWRP	Lower NB	Summer	2001	California corbina	3	57.8
SCCWRP	Lower NB	Summer	2001	California corbina	3	4.4
SCCWRP	Lower NB	Summer	2001	Diamond turbot	6	≤5
SCCWRP	Lower NB	Summer	2001	Spotfin croaker	6	≤5
SCCWRP	Lower NB	Summer	2001	Spotfin croaker	6	≤5
SCCWRP	Lower NB	Summer	2001	California halibut	4	≤5
SCCWRP	Upper NB	Winter	00/01	Black perch	3	≤5
SCCWRP	Upper NB	Winter	00/01	California halibut	4	≤5
SCCWRP	Upper NB	Winter	00/01	California halibut	4	≤5
SCCWRP	Upper NB	Winter	00/01	California halibut	4	≤5
SCCWRP	Upper NB	Winter	00/01	Diamond turbot	6	≤5
SCCWRP	Upper NB	Winter	00/01	Diamond turbot	6	≤5
SCCWRP	Upper NB	Winter	00/01	Diamond turbot	6	≤5
SCCWRP	Upper NB	Winter	00/01	Shiner perch	8	≤5
SCCWRP	Upper NB	Winter	00/01	Spotted sand bass	3	≤5
SCCWRP	Upper NB	Winter	00/01	Spotted turbot	4	≤5
SCCWRP	Upper NB	Summer	2001	Jacksmelt	6	≤5
SCCWRP	Upper NB	Summer	2001	Jacksmelt	6	9.9
SCCWRP	Upper NB	Summer	2001	Jacksmelt	6	≤5
SCCWRP	Upper NB	Summer	2001	Diamond turbot	6	≤5
SCCWRP	Upper NB	Summer	2001	Diamond turbot	6	≤5

Personal communication from Randall Yates (2006) at the California State Water Resources Control Board. Fish residue data spread sheets from the Coastal Fish Contamination Program

composites exceeded the guidance of 20 ppb. Of these nine fish composites, only two, representing one species, spotted sand bass, were from a species considered to be resident to Newport Bay. The remaining seven composites above 20 ppb were from coastal species, i.e., those that migrate up and down the coast. Coastal species could have received exposures from higher PCB levels outside of Newport Bay. Even the spotted sand bass migrates off shore in the winter, creating the possibility of exposure to PCBs outside of Newport Bay.

6.2 California Sport Fish Guidance to Protect Human Health

The SARWQCB (2006) report states that three resident species from upper Newport Bay displayed residue levels in excess of the guidance, whereas the underlying data cited in Appendix A-1 states that only one resident species, spotted sand bass, from upper Newport Bay, was in excess of the guidance. Review of the cited studies supports the finding of just one resident species, spotted sand bass, with residues above 20 ppb.

Another important issue is the State Board's TMDL listing policy. The policy requires 6 exceedances for 63 measurements for listing and the initiation of a TMDL (Table 2-3 in SARWQCB 2006). The policy also advises Regional Boards to not consider migratory (coastal) fish because the residues can come from other watersheds. This concept would be particularly true for Newport Bay because of coastal migration of fish from the very highly contaminated Palos Verdes Shelf, only 30 miles to the north. With only two composites from one species of resident fish in excess of the guidance, the listing criteria is not met, meaning that a TMDL for PCBs would not be recommended by the State Board for Newport Bay.

More important than meeting the State Board's listing policy for a TMDL, the health risks of PCB from ingestion of resident sport fish caught in Newport Bay can be shown to be insignificant. The average residue in six composites of spotted sand bass was 21 ppb. This average assumes PCB concentrations at one-half the detection limit of 5 ppb in composites where PCBs were not detected. All other residential sport fish species were below 20 ppb, with most being below the detection limit of 5 ppb. Since sport fishermen eat several different species of resident fish, the average residue of PCBs ingested will be well below the guidance of 20 ppb. The average PCB residue in all 63 resident and coastal sport fish composites is 12 ppb, a value below the guidance. In addition, the fish residue data are 11 years old or older, meaning that with no new sources of PCB input and further decay of residues, exposures are even lower today and will continue to decrease in the future.

7 Summary

DDT, chlordane, toxaphene and the PCBs are persistent organochlorines that are still found in aquatic environments of Newport Bay and Watershed (Orange County, California), decades after their use was discontinued. Under the Clean Water Act, organochlorines are regulated by a total maximum daily load (TMDL) to achieve levels that protect wildlife and human health. Stakeholders in the Newport Bay Watershed and an Independent Advisory Panel (IAP) requested by the Regional Board and administered by Orange County have questioned the quality of the science used to establish TMDL targets by US EPA Region IX and the Santa Ana Regional Water Quality Control Board. This review brings together a number of technical reports written by stakeholder consultants that address the scientific basis for the organochlorine TMDLs for Newport Bay and Watershed.

Urbanization of former agricultural lands has effectively capped soil organochlorine residues, reducing runoff into aquatic environments. Sediment controls have further reduced the movement of organochlorines from soil to aquatic environments. Residues in soil, water, sediments and biota are declining. For example, DDT in red shiner and mussels from Newport Bay and Watershed is declining exponentially with a half-life of 3.8 years and 5.2 years, respectively.

Sediment TMDL targets for total DDT, based on threshold effects levels (TELs), are inappropriate due to outdated, inaccurate and misinterpreted data. The TEL method ignores important dose-response relationships in the data sets used to calculate TELs. TELs for total DDT greatly overestimate thresholds for sediment toxicity, when compared to dose-response studies. The problem with using TELs as TMDL sediment targets is that risks are over-estimated, resulting in the assignment of resources disproportionate to risk and thereby not minimizing overall risk to humans and wildlife.

Terns, cormorants and several other avian species found in Newport Bay and Watershed are less sensitive to the reproductive effects of DDTs than ospreys and brown pelicans. Residues of DDE in eggs in excess of 10 ppm, resulting in eggshell thinning of 15% or more, appear to be necessary to produce significant hatching failure. The lack of a correlation between DDE levels in Forster's tern eggs and eggshell thickness indicates that reproduction in the closely related and threatened least tern probably will not be affected by DDT levels that currently exist in Newport Bay. The IAP has recommended the least tern as an indicator species for potential toxicity of DDT to wildlife in Newport Bay and Watershed.

The rare appearance of marine mammals in Newport Bay is unlikely to result in significant exposures to organochlorines. Worldwide, studies of marine mammals have disclosed a wide range of body burdens of DDT, but few if any impacts have been clearly delineated. With wildlife tissue levels clearly on the decline, impacts that might be identified in the future are unlikely.

Fish tissue targets to protect wildlife were adopted from the 1972 National Academy of Sciences recommendations. The 50 ppb target for DDT in marine fish is protective, but very extensive study since 1972 indicates that 150 ppb is also protective of sensitive avian species like the osprey. Successful osprey breeding began in Newport Bay in 2006 and has continued through 2013. The 150 ppb level in marine fish is also the basis for the national marine water criterion, a level that would be expected to protect the brown pelican. The 1,000 ppb target for DDT in fresh water fish is not protective and should be lowered to 150 ppb.

Two additional fish guidance reports were considered, but were not used as TMDL targets to protect wildlife. The fish guidance of 14 ppb by Environment Canada is highly protective from the assumption that minimal eggshell thinning is toxic and from having used an insensitive species to assess worst case ingestion rates. The fish guidance of 50 ppb by US EPA Region IX (Biological Technical Assistance Group, BTAG) relied on the same study in brown pelicans that is the basis for the national criterion for DDT in water (150 ppb in the fish diet). The BTAG guidance is threefold lower by having assumed a rapid equilibration of DDTs between dietary fish and brown pelican eggs, even though the underlying data indicates that equilibration may take several years.

The triad analysis, although one kind of weight-of-evidence analysis, is incomplete and flawed, when it was used to assess impairment of aquatic biota by chlordane in Newport Bay. Relying on the mere presence of chlordane, along with hundreds of other chemicals in toxic sediments, constitutes an incomplete weight-of-evidence analysis. The chlordane ERM is not a reliable measure of toxicity thresholds and should not be used in a weight-of-evidence analysis to assess impairment of aquatic biota. Most important, one should consider the results of dose-response bioassays.

The New York State sediment guidance for toxaphene was promulgated as a TMDL freshwater target for the Watershed. The guidance relies on an unreferenced K_{ow} that is 158-fold greater than the K_{ow} recommended by the regulatory agencies promulgating the TMDL target. If the TMDL target is calculated with the recommended K_{ow}, the existing load of toxaphene in the Watershed is below the TMDL target.

TMDL fish targets for DDT and PCBs to protect sport fishermen in Newport Bay are based on California EPA guidance that was intended only as a risk tool and not as a TMDL standard. Even so, average residues of DDT and PCBs in sport fish collected in 2000 and 2001 from Newport Bay were below the guidance. More recent California EPA guidance weighs cancer and noncancer risks against the benefits of consuming fish. That guidance is not to exceed 520 ppb DDT in sport fish consumed at the rate of three 8 oz filets per week. Similar guidance for PCBs is 21 ppb. California EPA has not issued a sport fish advisory for Newport Bay.

Residues of organochlorines in fish have declined to levels that pose no known hazard to humans or wildlife. The margin of safety will continue to increase as residues in fish continue to decline.

Acknowledgments The authors thank Sat Tamaribuchi for his leadership during a long and arduous regulatory proceeding. Thanks go to Doreen DiBiasio-Erwin for permissions and literature searches, and to Aaron Mead of Flow Science Incorporated for help with data, tables and figures. Thanks also go to Carla Navarro Woods, Reserve Manager, California Department of Fish and Game, and to Scott Thomas and Nancy Kenyon of the Audubon Society who provided osprey breeding data in Newport Bay and Watershed. The authors thank The Irvine Company for funding the technical analysis of the organochlorine TMDL issues and the writing of this review.

References

Agency for Toxic Substances and Disease Registry (1994) Toxicological profile for chlordane. US DHHS, Atlanta, GA
Agency for Toxic Substances and Disease Registry (1996) Toxicological profile for toxaphene. US DHHS, Atlanta, GA
Agency for Toxic Substances and Disease Registry (2002) Toxicological profile for DDT, DDE, and DDD. US DHHS, Atlanta, GA
Allen MJ, Groce AK, Diener D, Brown J, Steinert SA, Deets G, Noblet JA, Moore SL, Diehl D, Jarvis ET, Raco-Rands V, Thomas C, Ralph Y, Gartman R, Cadien D, Weisberg SB, Mikel T (2002) Southern California Bight (1998) Regional monitoring program: V. Demersal fishes and megabenthic invertebrates, Southern California Coastal Water Research Project. http://www.sccwrp.org/data/1998_bight_survey.html
Allen MJ, Diehl DW, Zeng EY (2004) Bioaccumulation of contaminants in recreational and forage fish in Newport Bay, California in 2000-2002. SCCWRP technical report 436. ftp://ftp.sccwrp.org/pub/download/PDFs/436_bioaccum.pdf

Allen MJ, Mason AZ, Gossett R, Diehl DW, Raco-Rands V, Schlenk D (2008) Assessment of food web transfer of organochlorine compounds and trace metals in fishes in Newport Bay, California. Southern California Coastal Water Research Project, Costa Mesa, CA

Ames PL, Mersereau GS (1964) Some factors in the decline of the osprey in Connecticut. Auk 81:173–185

Ames PL (1966) DDT residues in the eggs of the osprey in the north-eastern United States and their relation to nesting success. J Appl Ecol 3(Supplement):87–97

Anderson DW, Hickey JJ, Risebrough RW, Hughes DF, Christensen RE (1969) Significance of chlorinated hydrocarbon residues to breeding pelicans and cormorants. Can Field Natural 83:91–112

Anderson DW, Hickey JJ (1972) Eggshell changes in certain North American birds. Proc Intern Ornithol Congr 15:514–540

Anderson DW, Jehl JR, Risebrough RW, Woods LA, Deweese LR, Edgecomb WG (1975) Brown pelicans: improved reproduction off the Southern California Coast. Science 190:806–808

Anderson DW, Jurek RM, Keith JO (1977) The status of brown pelicans at Anacapa Island in 1975. Calif Fish Game 63(1):4–10

Anderson DW, Gress F (1983) Status of a northern population of California brown pelicans. Condor 85:79–88

Anderson JM, Bay SM, Thompson BE (1988) Characteristics and effects of contaminated sediments from southern California SCCWRP contribution No. C-297. Southern California Coastal Water Research Project, Long Beach, CA

Audet DJ, Scott DS, Wiemeyer SN (1992) Organochlorines and mercury in osprey eggs from the eastern United States. J Raptor Res 26:219–224

Bacon CE, Jarman WM, Costa DP (1992) Organochlorine and polychlorinated biphenyl levels in pinniped milk from the Arctic, the Antarctic, California and Australia. Chemosphere 24:779–791

Baird SF, Girard CF (1853) Descriptions of new species of fishes, collected by captains R.B. Marcy and Geo. B. McClellan in Arkansas. Proc Acad Nat Sci Phila, pp 390–392

Bay SM, Greenstein DJ (2003) Newport Bay and San Diego Creek - chemistry results for water, sediment, and suspended sediment, report to SARWQCB, March 14. http://www.waterboards.ca.gov/water_issues/programs/tmdl/records/region_8/2006/ref451.pdf

Bay SM, Greenstein DJ, Brown JS (2004) Newport Bay sediment toxicity studies. SCCWRP technical report 433. ftp://ftp.sccwrp.org/pub/download/PDFs/433_newportbay_sediment.pdf

Becker DS, Barrick RC, Read LB (1989) Evaluation of the AET approach for assessing contamination of marine sediments in California. Report No. 90-3WQ. PTI Environmental Services, Bellevue, WA

Bilger MD, Brightbill RA, Campbell HL (1999) Occurrence of organochlorine compounds in whole fish tissue from streams of the lower Susquehanna River Basin, Pennsylvania and Maryland, 1992. Water-Resources Investigations Report 99-4065. U. S. Geological Survey, Lemoyne, PA

Blasius ME, Goodmanlowe GD (2008) Contaminants still high in top-level carnivores in the Southern California Bight: levels of DDT and PCBs in resident and transient pinnipeds. Mar Pollut Bull 56:1973–1982

Blus LJ (1970) Measurements of brown pelican eggshells from Florida and South Carolina. Bioscience 20:867–869

Blus LJ, Heath RG, Gish CD, Belisle AA, Prouty RM (1971) Eggshell thinning in the brown pelican: implication of DDE. Bioscience 21:1213–1215

Blus LJ, Gish CD, Belisle AA, Prouty RM (1972a) Logarithmic relationship of DDE residues to eggshell thinning. Nature 235:376–377

Blus LJ, Gish CD, Belisle AA, Prouty RM (1972b) Further analysis of the logarithmic relationship of DDE residues to eggshell thinning. Nature 240:164–166

Blus LJ, Belisle AA, Prouty RM (1974a) Relations of the brown pelican to certain environmental pollutants. Pestic Monit J 7:181–194

Blus LJ, Neely BS, Belisle AA, Prouty RM (1974b) Organochlorine residues in brown pelican eggs: relation to reproductive success. Environ Pollut 7:81–91

Blus LJ, Neely BS, Lamont TG, Mulhern B (1977) Residues of organochlorines and heavy metals in tissues and eggs of brown pelicans. Pestic Monit J 11:40–53

Blus L, Cromartie E, McNease L, Joanen T (1979a) Brown pelican: population status, reproductive success, and organochlorine residues in Louisiana, 1971-1976. Bull Environ Contam Toxicol 12:128–135

Blus LJ, Lamont TG, Neely BS (1979b) Effects of organochlorine residues on eggshell thickness, reproduction, and population status of brown pelicans (*Pelecanus occidentalis*) in South Carolina and Florida, 1969-1976. Pestic Monit J 12:172–184

Blus LJ, Prouty RM (1979) Organochlorine pollutants and population status of least turns in South Carolina. Wilson Bull 91:62–71

Blus LJ, Stafford CJ (1980) Breeding biology and relation of pollutants to black skimmers and gull-billed terns in South Carolina. Fish and Wildlife Service, Special Scientific Report - Wildlife No. 230

Blus LJ (1982) Further interpretation of the relation of organochlorine residues in brown pelican eggs to reproductive success. Environ Pollut 28:15–33

Blus LJ (1984) DDE in bird's eggs: comparison of two methods for estimating critical levels. Wilson Bull 96:268–276

Blus LJ, Henney CJ, Stafford CJ, Grove RA (1987) Persistence of DDT and metabolites in wildlife from Washington State orchards. Arch Environ Contam Toxicol 16:467–476

Boellstorff DE, Ohlendorf HM, Anderson DW, O'Neill EJ, Keith JO, Prouty RM (1985) Organochlorine chemical residues in white pelicans and western grebes from the Klamath Basin, California. Arch Environ Contam Toxicol 14:485–493

Bolton HS, Breteler RJ, Vigon BW, Scanlon JA, Clark SL (1985) National perspective on sediment quality, Battelle, Washington Environmental Program Office. EPA, Washington, DC, Contract No. 68-01-6986

Borrell A, Cantos G, Pastor T, Aguilar A (2001) Organochlorine compounds in common dolphins (Delphinus delphis) from the Atlantic and Mediterranean waters of Spain. Environ Pollut 114:265–274

Borrell A, Aguilar A (2007) Organochlorine concentrations declined during 1987-2002 in western Mediterranean bottlenose dolphins, a coastal top predator. Chemosphere 66:347–352

Bredhult C, Backlin B-M, Bignert A, Olovsson M (2008) Study of the relation between the incidence of uterine leiomyomas and the concentrations of PCB and DDT in Baltic gray seals. Reprod Toxicol 25:247–255

Brodberg RK, Pollock GA (1999) Prevalence of selected target chemical contaminants in sport fish from two California lakes: public health designed screening study. Pesticide and Environmental Toxicology Section, Office of Environmental Health Hazard Assessment, California Environmental Protection Agency, Sacramento, CA

Buchman MF (1999) Screening quick reference tables, NOAA HAZMAT Report 99-1, Coastal Protection and Restoration Division, National Oceanic and Atmospheric Administration, Seattle, http://response.restoration.noaa.gov/cpr/sediment/squirt/squirt.pdf.

Burt WH, Grossenheider RP (1976) A field guide to the mammals, 3rd edn. Houghton Mifflin, Boston, MA

Byard JL (2011) Scientific commentary on DDT levels and eggshell thinning in bird eggs collected in 2004 and 2005 by CH2M Hill from Newport Bay and Watershed. Technical report prepared for The Irvine Company.

Byard JL (2012a) Scientific commentary on the biological effects data sets (BEDS) used to calculate ecological risk guidance for organochlorines in sediments from Newport Bay and Watershed. Technical report prepared for The Irvine Company.

Byard JL (2012b) Osprey reproduction in Newport Bay. Update with 2012 breeding results. Technical report prepared for The Irvine Company.

California Department of Toxic Substances Control (2008) Currently Recommended U.S. Environmental Protection Agency Region 9 Biological Technical Assistance Group

(BTAG) Mammalian and Avian Toxicity Reference Values (TRVs). http://www.dtsc.ca.gov/AssessingRisk/upload/Eco_Btag-mammal-bird-TRV-table.pdf

California Mussel Watch Program (1980–2000). http://www.swrcb.ca.gov/programs/smw/

CSWRCB (1998) California State Water Resources Control Board, Sediment chemistry, toxicity, and benthic community conditions in selected water bodies of the Santa Ana Region. Final Report, Bay Protection and Toxic Cleanup Program

California Toxic Substances Monitoring Program (1983–2002). http://www.swrcb.ca.gov/programs/smw/index.html

Callahan MA, Slimak MW, Gabel NW, May IP, Fowler CF, Freed JR, Jennings P, Durfee RL, Whitmore FC, Maestri B, Mabey WR, Holt BR, Gould C (1979) Water-related environmental fate of 129 priority pollutants. Vol I. Introduction and technical background, metals and inorganics, pesticides and PCBs. EPA-440/4-79-029a

Canadian Council of Ministers of the Environment (1999) Protocol for the derivation of canadian tissue residue guidelines for the protection of wildlife that consume aquatic biota. Canadian Council of Ministers of the Environment, Winnipeg, MB, pp 1–18

Cardwell RD, Foreman DG, Payne TR, Wilbur DJ (1977) Acute and chronic toxicity of chlordane to fish and invertebrates. EPA-600/3-77-019, NTIS PB267-544

Castrillon J, Gomez-Campos E, Aguilar A, Berdie L, Borrell A (2010) PCB and DDT levels do not appear to have enhanced the mortality of striped dolphins (*Stenella coeruleoalba*) in the 2007 Mediterranean epizootic. Chemosphere 81:459–463

Chapman PM, Dexter RN, Long ER (1987) Synoptic measures of sediment contamination, toxicity and infaunal community composition (the sediment quality triad) in San Francisco Bay. Mar Ecol Progr Ser 37:75–96

Clark KE, Stansley W, Niles LJ (2001) Changes in contaminant levels in New Jersey osprey eggs and prey, 1989 to 1998. Arch Environ Contam Toxicol 40:277–284

Connolly JP, Glaser D (2002) p, p'-DDE bioaccumulation in female sea lions of the California Channel Islands. Continen Shelf Res 22:1059–1078

Coulter MC, Risebrough RW (1973) Shell-thinning in eggs of the ashy petrel (*Oceanodroma homochroa*) from the Farallon Islands. Condor 75:254–255

Custer TW, Mitchell CA (1987) Organochlorine contaminants and reproductive success of black Skimmers in South Texas, 1984. J Field Ornith 58:480–489

Custer TW, Custer CM, Hines RK, Gutreuter S, Stromborg KL, Allen PD, Melancon MJ (1999) Organochlorine contaminants and reproductive success of double-crested cormorants from Green Bay, Wisconsin, USA. Environ Toxicol Chem 18:1209–1217

Das K, Vossen A, Tolley K, Vikingsson G, Thron K, Muller G, Baumgartner W, Siebert U (2006) Interfollicular fibrosis in the thyroid of the harbour porpoise: an endocrine disruption? Arch Environ Contam Toxicol 51:720–729

Debier CG, Ylitalo M, Weise M, Gulland F, Costa DB, Le Boeuf BJ, de Tillesse T, Larondelle Y (2005) PCBs and DDT in the serum of juvenile California sea lions: associations with vitamins A and E and thyroid hormones. Environ Pollut 134:323–332

de Bruijn J, Busser F, Seinen W, Hermens J (1989) Determination of octanol/water partition coefficients for hydrophobic organic chemicals with the "slow-stirring" method. Environ Toxicol Chem 8:499–512

Del Toro L, Heckel G, Camacho-Ibar VF, Schramm Y (2006) California sea lions (*Zalophus californianus californianus*) have lower chlorinated hydrocarbon contents in northern Baja California, Mexico, than in California, USA. Environ Pollut 142:83–92

Dirksen S, Boudewijn TJ, Slager LK, Mes RG, van Schaick MJM, de Voogt P (1995) Reduced breeding success of cormorants (*phalacrocorax carbo sinensis*) in relation to persistent organochlorine pollution of aquatic habitats in the Netherlands. Environ Pollut 88:119–132

Drescher HE, Harms U, Huschenbeth E (1977) Organochlorines and heavy metals in the harbour seal *Phoca vitulina* from the German North Sea Coast. Mar Biol 41:99–106

Duinker JC, Hillebrand MTJ, Nolting RF (1979) Organochlorines and metals in harbour seals (Dutch Wadden Sea). Mar Pollut Bull 10:360–364

Eisler R (1990) Chlordane hazards to fish, wildlife, and invertebrates: a synoptic review. U.S. Fish and Wildlife Service, Biological Report 85(1.21), NTIS PB91-114223

Elfes CT, VanBlaricom GR, Boyd D, Calambokidis J, Clapham PJ, Pearce RW, Robbins J, Salinas JC, Straley JM, Wade PR, Krahn MM (2010) Geographic variation of persistent organic pollutant levels in humpback whale (*Megaptera novaeangliae*) feeding areas of the North Pacific and North Atlantic. Environ Toxicol Chem 29:824–834

Elliot JE, Noble DG, Norstrom RJ, Whitehead PE (1989) Organochlorine contaminants in seabird eggs from the pacific coast of Canada, 1971-1986. Environ Monit Assess 12:67–82

Elliott JE, Machmer MM, Wilson LK, Henny CJ (2000) Contaminants in ospreys from the Pacific Northwest: II. Organochlorine pesticides, polychlorinated biphenyls, and mercury, 1991-1997. Arch Environ Contam Toxicol 38:93–106

Environment Canada (1992) Interim criteria for quality assessment of St. Lawrence River sediment. ISBN 0-662-19849-2, St. Lawrence Action Plan, St. Lawrence Centre and Ministère de l'Environnement du Québec.

Environment Canada (2000) Environmental quality assessments for PCBs, DDT and toxaphene, vol 5, Monograph series. Canadian Association on Water Quality, Ottawa, ON, pp 1–178

Ewins PJ (1997) Osprey (*Pandion haliaetus*) populations in forested areas of North America: changes, their causes and management recommendations. J Raptor Res 31:138–150

Ewins PJ, Postupalsky S, Hughes KD, Weseloh DV (1999) Organochlorine contaminant residues and shell thickness of eggs from known-age female ospreys (*Pandion haliaetus*) in Michigan during the 1980s. Environ Pollut 104:295–304

Faber RA, Hickey JJ (1973) Eggshell thinning, chlorinated hydrocarbons, and mercury in inland aquatic bird eggs, 1969 and 1970. Pestic Mon J 7:27–36

Fair PA, Adams J, Mitchum G, Hulsey TC, Reif JS, Houde M, Muir D, Wirth E, Wetzel D, Zolman E, McFee W, Bossart GD (2010) Contaminant blubber burdens in Atlantic bottlenose dolphins (*Tursiops truncatus*) from two southeastern US estuarine areas: concentrations and patterns of PCBs, pesticides, PBDEs, PFCs and PAHs. Sci Total Environ 408:1577–1597

Falkenberg ID, Dennis TE, Williams BD (1994) Organochlorine pesticide contamination in three species of raptor and their prey in South Australia. Wildl Res 21:163–173

Fillman G, Hermanns L, Fileman TW, Readman JW (2007) Accumulation patterns of organochlorines in juveniles of *Arctocephalus australis* found stranded along the coast of Southern Brazil. Environ Pollut 146:262–267

Findholt SL, Trost CH (1985) Organochlorine pollutants, eggshell thickness, and reproductive success of black-crowned night-herons in Idaho, 1979. Colon Waterb 8:32–41

Flow Science Inc, Byard JL, Tjeerdema RS, QEA Environmental Consultants Inc (2006) DDT analysis for the Newport Bay Watershed. http://www.waterboards.ca.gov/santaana/html/newport_oc_tmdl.html

Fossi C, Focardi S, Leonzio C, Renzoni A (1984) Trace-metals and chlorinated hydrocarbons in bird's eggs from the delta of the Danube. Environ Conser 11:345–350

Fox GA (1976) Eggshell quality: its ecological and physiological significance in a DDE-contaminated common tern population. Wilson Bull 88:459–477

Gamble LR, Blankinship DR, Jackson GA (1987) Contaminants in brown pelican eggs collected from Texas and Mexico, 1986. U.S. Fish and Wildlife Service, No. R2-87-01.

Goodbred SL, Ledig DB, Roberts CA (1996) Organochlorine contamination in eggs, prey and habitat of the light-footed clapper rail in three southern California marshes. US Fish and Wildlife Service, Carlsbad, CA

Greene EP, Greene AE, Freedman B (1983) Foraging behavior and prey selection by ospreys in coastal habitats in Nova Scotia, Canada. In: Bird D, Seymour N, Gerrard J (eds) Biology and management of bald eagles and ospreys. McGill University and Raptor Research Foundation, Montreal, QC, pp 257–267

Greig DJ, Ylitalo GM, Wheeler EA, Boyd D, Gulland FMD, Yanagida GK, Harvey JT, Hall AJ (2011) Geography and stage of development affect persistent organic pollutants in stranded and wild-caught harbor seal pups from central California. Sci Total Environ 409:3537–3547

Gress F, Risebrough RW, Anderson DW, Kiff LF, Jehl JR (1973) Reproductive failures of double-crested cormorants in Southern California and Baja California. Wilson Bull 85:197–208

Gress F (1995) Organochlorines, eggshell thinning, and productivity relationships in brown pelicans breeding in the Southern California Bight. Ph.D. Thesis, University of California, Davis

Hall AJ, Law RJ, Wells DE, Harwood J, Ross HM, Kennedy S, Allchin CR, Campbell LA, Pomeroy PP (1992) Organochlorine levels in common seals (*Phoca vitulina*) which were victims and survivors of the 1988 phocine distemper epizootic. Sci Total Environ 115:145–162

Hall AJ, Thomas GO, McConnell BJ (2009) Exposure to persistent organic pollutants and first-year survival probability in gray seal pups. Environ Sci Technol 43:6364–6369

Hart DR, Fitchko J, McKee PM (1988) Development of sediment quality guidelines. Phase II guideline development. BEAK Ref. 2437.1, Prepared by BEAK Consultants Limited, Brampton, Ontario. Prepared for Ontario Ministry of the Environment, Toronto, ON.

Hays H, Risebrough RW (1972) Pollutant concentrations in abnormal young terns from Long Island Sound. Auk 89:19–35

Hayteas DL, Duffield DA (1997) The determination by HPLC of p, p'-DDE residues in marine mammals stranded on the Oregon Coast, 1991–1995. Mar Pollut Bull 34:844–848

Hazeltine W (1972) Disagreements on why brown pelican eggs are thin. Nature 239:410–411

Henderson C, Inglis A, Johnson WL (1971) Organochlorine insecticide residues in fish – fall 1969 National Pesticide Monitoring Program. Pestic Monit J 5:1–11

Henny CJ, Ogden JC (1970) Estimated status of osprey populations in the United States. J Wildl Manag 34:214–217

Henny CJ, Byrd MA, Jacobs JA, McLain PD, Todd MR, Halla BF (1977) Mid-Atlantic coast osprey populations: present numbers, productivity, pollutant contamination, and status. J Wildl Manag 41:254–265

Henny CJ, Ward FP, Riddle KE, Prouty RM (1982a) Migratory peregrine falcons, *Falco peregrinus*, accumulate pesticides in Latin America during winter. Can Field Natural 96:333–338

Henny CJ, Blus LJ, Prouty RM (1982b) Organochlorine residues and shell thinning in Oregon seabird eggs. Murrelet 63:15–21

Henny CJ (1983) Distribution and abundance of nesting ospreys in the United States. In: Bird D, Seymour N, Gerrard J (eds) Biology and management of bald eagles and ospreys. McGill University and Raptor Research Foundation, Montreal, QC, pp 175–186

Henny CJ, Blus LJ, Hulse CS (1985) Trends and effects of organochlorine residues on Oregon and Nevada wading birds, 1979-83. Colon Waterb 8:117–128

Henny CJ, Kaiser JL, Grove RA, Bentley VR, Elliott JE (2003) Biomagnification factors (fish to osprey eggs from Willamette River, Oregon, U.S.A.) for PCDDs, PCDFs, PCBs and OC pesticides. Environ Monit Assess 84:275–315

Henny CJ, Grove RA, Kaiser JL, Bentley VR (2004) An evaluation of osprey eggs to determine spatial residue patterns and effects of contaminants along the lower Columbia River, U.S.A. In: Chancellor R, Meyburg B (eds) Raptors Worldwide. WWGBP/MME, pp 369-388.

Henny CJ, Anderson TW, Crayon JJ (2008) Organochlorine pesticides, polychlorinated biphenyls, metals, and trace elements in waterbird eggs, Salton Sea, California, 2004. Hydrobiologia 604:137–149

Henry J, Best PB (1983) Organochlorine residues in whales landed at Durban, South Africa. Mar Pollut Bull 14:223–227

Hobbs KE, Muir DCG, Born EW, Dietz R, Haug T, Metcalfe T, Metcalfe C, Oien N (2003) Levels and patterns of persistent organochlorines in Minke whale (*Balaenoptera acutorostrata*) stocks from the North Atlantic and European Arctic. Environ Pollut 121:239–252

Hoffman DJ, Smith GJ, Rattner BA (1993) Biomarkers of contaminant exposure in common terns and black-crowned night herons in the Great Lakes. Environ Toxicol Chem 12:1095–1103

Hoguet J, Keller JM, Reiner JL, Kucklick JR, Bryan CE, Moors AJ, Pugh RS, Becker PR (2013) Spatial and temporal trends of persistent organic pollutants and mercury in beluga whales (*Delphinapterus leucas*) from Alaska. Sci Total Environ 449:285–294

Holsapple MP, Pitot HC, Cohen SH, Boobis AR, Klaunig JE, Pastoor T, Dellarco VL, Dragan YP (2006) Mode of action in relevance of rodent liver tumors to human cancer risk. Toxicol Sci 89:51–56

Hornsby AG, Wauchope RE, Herner AE (1996) Pesticide properties in the environment. Springer, New York, NY, pp 65–66

Hothem RL, Zador SG (1995) Environmental contaminants in eggs of California least terns (*Sterna antillarum browni*). Bull Environ Contam Toxicol 55:658–665

Hothem RL, Powell AN (2000) Contaminants in eggs of western snowy plovers and California least terns: is there a link to population decline? Bull Environ Contam Toxicol 65:42–50

Hunt EG (1969) Pesticides as possible factors affecting bird populations. Chapter 41, Pesticide residues in fish and wildlife of California. In: Joseph J (ed) Peregrine falcon populations, their biology and decline, hickey. University of Wisconsin Press, Madison, WI, pp 455–460

IAP (2009) Final report of the Orange County independent advisory panel for the assessment of TMDL targets for organochlorine compounds for the Newport Bay, August 4th.

IEPA (1988) An intensive survey of the DuPage River Basin, 1983, IEPA/WPC/88-010. Division of Water Pollution Control, Springfield, IL

Ingles LG (1965) Mammals of the pacific states. Stanford Univ Press, Stanford, CA

Jehl JR (1973) Studies of a declining population of brown pelicans in Northwestern Baja California. Condor 75:69–79

Johnson DR, Melquist WE, Schroeder GJ (1975) DDT and PCB levels in Lake Coeur d'Alene, Idaho, osprey eggs. Bull Environ Contam Toxicol 13:401–405

JRB Associates (1984) Background and review document of the development of sediment criteria. EPA Contract No. 68-01-6388, JRB Project No. 2-813-03-852-84, Washington, DC

Kajiwara N, Kannan K, Muraoka M, Watanabe M, Takahashi S, Gulland F, Olsen H, Blankenship AL, Jones PD, Tanabe S, Giesy JP (2001) Organochlorine pesticides, polychlorinated biphenyls, and butyltin compounds in blubber and livers of stranded California sea lions, elephant seals, and harbor seals from coastal California, USA. Arch Environ Contam Toxicol 41:90–99

Kannan K, Kajiwara N, Le Boeuf BJ, Tanabe S (2004) Organochlorine pesticides and polychlorinated biphenyls in California sea lions. Environ Pollut 131:425–434

Keith JO (1969) Variations in the biological vulnerability of birds to insecticides. In: Gillett J (ed) The biological impact of pesticides in the environment. Oregon State University, Corvallis, OR, pp 36–39

Keith JO, Woods LA, Hunt EG (1970) Reproductive failure in brown pelicans on the pacific coast. Trans North Am Wildl Nat Res Conf 35:56–63

Keith JA, Gruchy IM (1972) Residue levels of chemical pollutants in North American birdlife. Proc Intern Ornithol Congr 15:437–454

Kerr R (2006) Successful osprey nesting at Back Bay, Tracks, a publication of the Newport Bay Naturalists & Friends, p 1, September–November

King KA, Flickenger EL, Hildebrand HH (1977) The decline of brown pelicans on the Louisiana and Texas Gulf Coast. Southwest Natural 21:417–431

King KA, Flickenger EL, Hildebrand HH (1978) Shell thinning and pesticide residues in Texas aquatic bird eggs, 1970. Pestic Monit J 12:16–21

King KA, Blankinship DR, Payne E, Krynitsky AJ, Hensler GL (1985) Brown pelican populations and pollutants in Texas 1975-1981. Wilson Bull 97:201–214

King KA, Krynitsky AJ (1986) Population trends, reproductive success, and organochlorine chemical contaminants in waterbirds nesting in Galveston Bay, Texas. Arch Environ Contam Toxicol 15:367–376

King KA, Custer TW, Quinn JS (1991) Effects of mercury, selenium, and organochlorine contaminants on reproduction of Forster's terns and black skimmers nesting in a contaminated Texas bay. Arch Environ Contam Toxicol 20:32–40

Klasing S, Brodberg R (2008) Development of fish contamination goals and advisory tissue levels for common contaminants in California sport fish: chlordane, DDTs, dieldrin, methylmercury, PCBs, selenium, and toxaphene. Office of Environmental Health Hazard Assessment, EPA, San Francisco, CA

Klevaine L, Skaare JU (1998) Organochlorine contaminants in northeast Atlantic Minke whales (*Balaenoptera acutorostrata*). Environ Pollut 101:231–239

Koeman JH, van Genderen H (1966) Some preliminary notes on residues of chlorinated hydrocarbon insecticides in birds and mammals in the Netherlands. J Appl Ecol 3:99–106

Koeman JH, Peters WHM, Smit CJ, Tjoie PS, de Ggoeij JJM (1972) Persistent chemical in marine mammals. T N O Nieuws 27:570–578

Lailson-Brito J, Dorneles PR, Azevedo-Silva CE, Azevedo AF, Vidal LG, Marigo J, Bertozzi C, Zanelatto RC, Bisi TL, Malm O, Torres JPM (2011) Organochlorine concentrations in Franciscana dolphins, *Pontoporia blainvillei*, from Brazilian waters. Chemosphere 84:882–887

Lailson-Brito J, Dorneles PR, Azevedo-Silva CE, Bisi TL, Vidal LG, Legat LN, Azevedo AF, Torres JPM, Malm O (2012) Organochlorine compound accumulation in delphinids from Rio de Janeiro State, southeastern Brazilian coast. Sci Total Environ 433:123–131

Law RJ, Bersuder P, Barry J, Barber J, Deaville R, Barnett J, Jepson PD (2013) Organochlorine pesticides and chlorobiphenyls in the blubber of bycaught female common dolphins from England and Wales from 1992-2006. Mar Poll Bull 69:238–242

Le Boeuf BJ, Bonnell ML (1971) DDT in California sea lions. Nature 234:108–110

Le Boeuf BJ, Giesy JP, Kannan K, Kajiwara N, Tanabe S, Debier C (2002) Organochlorine pollutants in California sea lions revisited. BMC Ecol 2:11–19

Lee GF, Taylor S (2001) Results of aquatic toxicity testing conducted during 1997-2000 within the Upper Newport Bay, Orange County, CA watershed. Report of G. Fred Lee & Associates. G. Fred Lee & Associates, El Marcero, CA

Lichtenstein E, Schultz K (1959) Persistence of some chlorinated hydrocarbon insecticide as influenced by soil types, rate of application and temperature. J Econ Entomol 52:124–131

Lieberg-Clark P, Bacon CE, Burns SA, Jarman WM, Le Boeuf BJ (1995) DDT in California sea lions: a follow-up study after 20 years. Mar Pollut Bull 30:744–745

Lincer JL (1975) DDE-induced eggshell-thinning in the American kestrel: a comparison of the field situation and laboratory results. J Appl Ecol 12:781–793

Lindvall ML, Low JB (1980) Effects of DDE, TDE, and PCBs on shell thickness of western grebe eggs, Bear River Migratory Bird Refuge, Utah - 1973-1974. Pestic Monit J 14:108–111

Litz JA, Garrison LP, Fieber LA, Martinez A, Contillo JP, Kucklick JR (2007) Fine-scale spatial variation of persistent organic pollutants in bottlenose dolphins (*Tursiops truncatus*) in Biscayne Bay, Florida. Environ Sci Technol 41:7222–7228

Long ER, Morgan LG (1990) The potential for biological effects of sediment-sorbed contaminants tested in the national status and trends program. NOAA technical memorandum NOS OMA 52, Seattle, Washington.

Long ER, MacDonald DD, Smith SL, Calder FD (1995) Incidence of adverse biological effects within ranges of chemical concentrations in marine and estuarine sediments. Environ Manage 19:81–97

Longcore JR, Stendell RC (1977) Shell thinning and reproductive impairment in Black Ducks after cessation of DDE dosage. Arch Environ Contam Toxicol 6:293–304

Luckas B, Vetter W, Fischer P, Heidemann G, Plotz J (1990) Characteristic chlorinated hydrocarbon patterns in the blubber of seals from different marine regions. Chemosphere 21:13–19

Lyman WJ, Glazer AE, Ong JH, Coons SF (1987) An overview of sediment quality in the United States. Final Report. US EPA Region V Contract No. 68-01-6951, Task 20, PB88-251384, Washington, DC.

MacCarter DL, MacCarter DS (1979) Ten-year nesting status of ospreys at Flathead Lake, Montana. Murrelet 60:42–49

MacDonald DD (1994) Sediment injury in the Southern California Bight: review of the toxic effects of DDTs and PCBs in sediments. Technical report prepared for the National Oceanic and Atmospheric Administration.

MacDonald DD, Carr RS, Calder FD, Long ER, Ingersoll CG (1996) Development and evaluation of sediment quality guidelines for Florida coastal waters. Ecotoxicology 5:253–278

Marking LL, Dawson VK, Allen JL, Bills TD, Rach JJ (1981) Biological activity and chemical characteristics of dredge material from ten sites on the Upper Mississippi River. Cooperative Agreement No. 14-16-0009-79-1020, U.S. Fish and Wildlife Service, La Crosse, Wisconsin, Supported by the U.S. Army Corps of Engineers, St. Paul District, Minnesota.

Martell MS, Henny CJ, Nye PE, Solensky MJ (2001) Fall migration routes, timing, and wintering sites of North American ospreys as determined by satellite telemetry. Condor 103:715–724

Martin PA, de Solla SR, Ewins P (2003) Chlorinated hydrocarbon contamination in osprey eggs and nestlings from the Canadian Great Lakes Basin, 1991-1995. Ecotoxicology 12: 209–224

Masters P, Inman D (2000) Transport and Fate of Organochlorines discharged to the salt marsh at Upper Newport Bay, California, USA. Environ Toxicol Chem 19:2076–2084

McKinney MA, De Guise S, Martineau D, Beland P, Lebeuf M, Letcher RJ (2006) Organohalogen contaminants and metabolites in beluga whale (*Delphinapterus leucas*) liver from two Canadian populations. Environ Toxicol Chem 25:1246–1257

McLeese DW, Metcalfe CD (1980) Toxicities of eight organochlorine compounds in sediment and seawater to *Crangon septemspinosa*. Bull Environ Contam Toxicol 25:921–928

McLeese DW, Burridge LE, Van Dinter J (1982) Toxicities of five organochlorine compounds in water and sediment to *Nereis virens*. Bull Environ Contam Toxicol 28:216–220

Mendenhall VM, Prouty RM (1978) Recovery of breeding success in a population of brown pelicans. Proc Colon Waterb Group 2:65–70

Mischke T, Brunetti K, Acosta V, Weaver D, Brown M (1985) Agricultural sources of DDT residues in California's environment. A report prepared in response to House Resolution No. 53 (1984). Environmental Hazards Assessment Program, Dept. of Food and Agriculture, Sacramento, CA

Mora MA (1997) Transboundary pollution: persistent organochlorine pesticides in migrant birds of the Southwestern United States and Mexico. Environ Toxicol Chem 16:3–11

Morrison ML, Slack RD, Shanley E (1978) Declines in environmental pollutants in olivaceous cormorant eggs from Texas, 1970-77. Wilson Bull 90:640–642

Morrison ML, Kiff LF (1979) Eggshell thickness in American shorebirds before and since DDT. Can Field Natural 93:187–190

Mossner S, Ballschmiter K (1997) Marine mammals as global indicators for organochlorines. Chemosphere 34:1285–1296

Nash RG, Woolson EA (1967) Persistence of chlorinated hydrocarbon insecticides in soils. Science 157:924–927

NAS (1972) Water quality criteria. A report of the Committee on Water Quality Criteria. Environmental Studies Board, Washington, DC

Neff JM, Bean DJ, Cornaby BW, Vaga RM, Gulbransen TC, Scalon JA (1986) Sediment quality criteria methodology validation. Calculation of screening level concentrations from field data. Prepared for U.S. EPA, Criteria and Standards Division by Battelle, Washington Environmental Program Office, Washington, DC.

New York State Department of Environmental Conservation (1998) Technical guidance for screening contaminated sediments. Division of Fish, Wildlife and Marine Resources, New York, NY

Newport Bay Conservancy (2013) http://newportbay.org/projects/army-corps-of-engineers-project/

Newton I, Bogan J (1978) The role of different organo-chlorine compounds in the breeding of British sparrowhawks. J Appl Ecol 15:105–116

Niimi S, Watanabe MX, Kim EY, Iwata H, Yasunaga G, Fujise Y, Tanabe S (2005) Molecular cloning and mRNA expression of cytochrome P4501A1 and 1A2 in the liver of common Minke whales (*Balaenoptera acutorostrata*). Mar Pollut Bull 51:784–793

Nino-Torres CA, Gardner SC, Zenteno-Savin T, Ylitalo GM (2009) Organochlorine pesticides and polychlorinated biphenyls in California sea lions (*Zalophus californianus californianus*) from the Gulf of California, Mexico. Arch Environ Contam Toxicol 56:350–359

Nino-Torres CA, Zenteno-Savin T, Gardner SC, Urban-Ramirez J (2010) Organochlorine pesticides and polychlorinated biphenyls in fin whales (*Balaenoptera physalus*) from the Gulf of California. Environ Toxicol 25:381–390

Nisbet ICT, Reynolds LM (1984) Organochorine residues in common terns and associated estuarine organisms, Massachusetts, USA, 1971-1981. Mar Environ Res 11:33–66

Noble DG, Elliott JE (1990) Levels of contaminants in Canadian raptors, 1966 to 1988, effects and temporal trends. Can Field Natural 104:222–243

Ohlendorf HM, Schaffner FC, Custer TW, Stafford CJ (1985) Reproduction and organochlorine contaminants in terns in San Diego Bay. Colon Waterb 8:42–53

Ohlendorf HM, Custer TW, Lowe RW, Rigney M, Cromartie E (1988) Organochlorines and mercury in eggs of coastal terns and herons in California, USA. Colon Waterb 11:85–94

Orange County Public Facilities and Resources Department (1980–86) Unpublished sediment data provided by Bruce Moore

Orange County Watersheds (2013) Unified Program Effectiveness Assessments (PEA), 2006-2011, Section C-11. Santa Ana region water quality. Monitoring summary & analyses. http://ocwatersheds.com/documents/damp/pea/2011_12_orange_county_stormwater_program_unified_pea

O'Shea TJ, Brownell RL, Clark DR, Walker WA, Gay ML, Lamont TG (1980) Fish, wildlife and estuaries: organochlorine pollutants in small cetaceans from the Pacific and south Atlantic Oceans, Nov. 1968 – June 1976. Mar Pollut Bull 36:159–164

O'Shea TJ, Brownell RL (1996) California sea lion (*Zalophus californianus*) populations and ΣDDT contamination. Mar Pollut Bull 36:159–164

Pavlou S, Kadeg R, Turner A, Marchlik M (1987) Sediment quality criteria methodology validation: uncertainty analysis of sediment normalization theory for nonpolar organic contaminants. Battelle Washington Environmental Program Office, Washington, DC

Peakall DB, Lincer JL, Risebrough RW, Pritchard JB, Kinter WB (1973) DDE-induced egg-shell thinning: structural and physiological effects in three species. Comp Gen Pharmacol 4:305–313

Peakall DB (1975) Physiological effects of chlorinated hydrocarbons on avian species. Environ Sci Res 6:343–360

Pearce PA, Peakall DB, Reynolds LM (1979) Shell thinning and residues of organochlorines and mercury in seabird eggs, eastern Canada, 1970-76. Pestic Monit J 13:61–68

Persaud D, Jaagumagi R, Hayton A (1991) The provincial sediment quality guidelines, Draft. Water Resources Branch, Ontario Ministry of the Environment, Toronto, ON

Peterson RT (1969) Population trends of ospreys in the northeastern United States. In: Hickey J (ed) Peregrine falcon populations, their biology and decline. University of Wisconsin Press, Madison, WI, pp 333–343

Pitot HC, Dragan YP (1996) Chemical carcinogenesis. In: Klassen C (ed) Casarett & Doull's toxicology, the basic science of poisons, 5th edn. McGraw-Hill, New York, NY, pp 238–239

Pollock GA, Uhas IJ, Fan AM, Wisniewski JA, Witherell I (1991) A study of chemical contamination of marine fish from Southern California. II. Comprehensive study. Office of Environmental Health Hazard Assessment, California Environmental Protection Agency, Sacramento, CA

Poole AF (1989) Ospreys a natural and unnatural history. Cambridge Press, New York, NY

Porter RD, Wiemeyer SN (1969) Dieldrin and DDT: effects on sparrow hawk eggshells and reproduction. Science 165:199–200

Racke K, Skidmore M, Hamilton D (1997) Pesticide fate in tropical soils. Pure Appl Chem 68:1349–1371

Ramsdell JS (2010) Neurological disease rises from ocean to bring model for human epilepsy to life. Toxins 2:1646–1675

Rattner BA, McGowan PC, Golden NH, Hatfield JS, Toschik PC, Lukei RF, Hale RC, Schmitz-Afonso I, Rice CP (2004) Contaminant exposure and reproductive success of ospreys (*Pandion haliaetus*) nesting in Chesapeake Bay regions of concern. Arch Environ Contam Toxicol 47:126–140

Reed S (2010) Osprey chick hatches at San Joaquin Marsh nesting platform. Irvine Ranch Water District news release, July 22

Reese JG (1977) Reproductive success of ospreys in central Chesapeake Bay. Auk 94:202–221

Reicher M (2010) Ospreys make a comeback in Orange County. Los Angeles Times, Greenspace, pp 1-2, September 13

Reijnders PJH (1980) Organochlorine and heavy metal residues in harbour seals from the Wadden Sea and their possible effects on reproduction. Neth J Sea Res 14:30–65

Risebrough RW, Menzel DB, Martin DJ, Olcott HS (1967) DDT residues in Pacific sea birds: a persistent insecticide in marine food chains. Nature 216:589–591

Risebrough RW, Davis J, Anderson DW (1969) Effects of various chlorinated hydrocarbons. In: Gillett J (ed) The biological impact of pesticides in the environment. Oregon State University, Corvallis, OR, pp 40–53

Risebrough RW, Sibley FC, Kirven MN (1971) Reproductive failure of the brown pelican on Anacapa Island in 1969. Am Bird 25:8–9

Risebrough RW (1972) Reply to Hazeltine. Nature 240:164

Roos AM, Backlin B-MVM, Helander BO, Riget FF, Eriksson UC (2012) Improved reproductive success in otters (*Lutra lutra*), grey seals (*Halichoerus grypus*) and sea eagles (*Haliaeetus albicilla*) from Sweden in relation to concentrations of organochlorine contaminants. Environ Pollut 170:268–275

Routti H, Nyman M, Jenssen BM, Backman C, Koistinen J, Gabrielsen GW (2008) Bone-related effects of contaminants in seals may be associated with vitamin D and thyroid hormones. Environ Toxicol Chem 27:873–880

Sakai H, Iwata H, Kim E-Y, Tsydenova O, Miyazaki N, Petrov EA, Batoev VB, Tanabe S (2006) Constitutive androstane receptor (CAR) as a potential sensing biomarker of persistent organic pollutants (POPs) in aquatic mammal: molecular characterization, expression level, and ligand profiling in Baikal seal (*Pusa sibirica*). Toxicol Sci 94:57–70

SARWQCB (1998) TMDL for sediment in Newport Bay

SARWQCB (1999) Fecal coliform TMDL

SARWQCB (2006) Total maximum daily loads for organochlorine compounds in San Diego Creek, Upper and Lower Newport Bay, Orange County, California. http://www.waterboards.ca.gov/santaana/html/newport_oc_tmdl.html

SCCWRP (1998) Sediment DDT measurements downloaded from bight 98 survey. http://www.sccwrp.org/data/1998_bight_survey.html

Schafer HA, Gossett RW, Ward CF, Westcott AM (1984) Chlorinated hydrocarbons in marine mammals. In: Bascom W (ed) Biennial report. Southern California Coastal Water Research Project, Long Beach, CA, pp 1983–1984

Schmitt CJ, Zajicek JL, Peterman PH (1990) National contaminant biomonitoring program: residues of organochlorine chemicals in U. S. freshwater fish, 1976-1984. Arch Environ Contam Toxicol 19:748–781

Schmitt CJ, Ludke JL, Walsh D (1981) Organochlorine residues in fish; 1970-1974: National Pesticide Monitoring Program. Pestic Monit J 14:136–206

Schreiber RW, Risebrough RW (1972) Studies of the brown pelican. Wilson Bull 84:119–135

Schuytema GS, Nebecker AV, Griffis WL, Miller CE (1989) Effects of freezing on toxicity of sediments contaminated with DDT and endrin. Environ Toxicol Chem 8:883–891

Setmire JG, Schroeder RA, Densmore JN, Goodbred SL, Audet DJ, Radke WR (1993) Detailed study of water quality, bottom sediment, and biota associated with irrigation drainage in the Salton Sea area, California, 1988–1990. Water-Resources Investigations Report 93-4014. U. S. Geological Survey, Sacramento, CA

Shaw SD, Brenner D, Bourakovsky A, Mahaffey CA, Perkins CR (2005) Polychlorinated biphenyls and chlorinated pesticides in harbor seals (*Phoca vitulina concolor*) from the northwestern Atlantic coast. Mar Pollut Bull 50:1069–1084

Shoham-Frider E, Kress N, Wynne D, Scheinin A, Roditi-Elsar M, Kerem D (2009) Persistent organochlorine pollutants and heavy metals in tissues of common bottlenose dolphin (*Tursiops truncatus*) from the Levantine Basin of the Eastern Mediterranean. Chemosphere 77:621–627

Siebert U, Pozniak B, Hansen KA, Nordstrom G, Teilmann J, van Elk N, Vossen A, Dietz R (2011) Investigations of thyroid and stress hormones in free-ranging and captive harbor porpoises (*Phocoena phocoena*): a pilot study. Aquat Mamm 37:443–453

Smith SL, MacDonald DD, Keenleyside KA, Ingersoll CG, Field LJ (1996) A preliminary evaluation of sediment quality assessment values for freshwater ecosystems. J Great Lakes Res 22:624–638

Smyth M, Berrow S, Nixon E, Rogan E (2000) Polychlorinated biphenyls and organochlorines in by-caught harbour porpoises *Phocoena phocoena* and common dolphins *Delphinus delphis* from Irish coastal waters. Proc Roy Ir Acad 100B:85–96

Spitzer PR, Risebrough RW, Walker W, Hernandez R, Poole A, Puleston D, Nisbet IC (1978) Productivity of ospreys in Connecticut – Long Island increases as DDE residues decline. Science 202:333–335

Spitzer PR, Poole AF (1980) Coastal ospreys between New York City and Boston: a decade of reproductive recovery 1969-1979. Am Bird 34:234–241

Spitzer PR, Poole AF, Scheibel M (1983) Initial population recovery of breeding ospreys in the region between New York City and Boston. In: Bird D, Seymour N, Gerrard J (eds) Biology and management of bald eagles and ospreys. McGill University and Raptor Research Foundation, Montreal, QC, pp 231–241

Steidl RJ, Griffin CR, Niles LJ (1991a) Differential reproductive success of ospreys in New Jersey. J Wildl Manag 55:266–272

Steidl RJ, Griffin CR, Niles LJ (1991b) Contaminant levels of osprey eggs and prey reflect regional differences in reproductive success. J Wildl Manag 55:601–608

Stewart D, Chisholm D (1971) Long-term persistence of BHC, DDT and chlordane in sandy loam soil. Can J Soil Sci 51:379–383

Stockin KA, Law RJ, Duignan PJ, Jones GW, Porter L, Mirimin L, Meynier L, Orams MB (2007) Trace elements, PCBs and organochlorine pesticides in New Zealand common dolphins (*Delphinus* sp.). Sci Total Environ 387:333–345

Stockin KA, Law RJ, Roe WD, Meynier L, Martinez E, Duignan PJ, Bridgen P, Jones B (2010) PCBs and organochlorine pesticides in Hector's (*Cephalorhynchus hectori hectori*) and Maui's (*Cephalorhynchus hectori maui*) dolphins. Mar Poll Bull 60:834–842

Struger J, Weseloh DV (1985) Great Lakes Caspian terns: egg contaminants and biological implications. Colon Waterb 8:142–149

Sutula M, Bay S, Santolo G, Zembal R (2005) Organochlorine and trace metal contaminants in the food web of the light-footed clapper rail, Upper Newport Bay, California. Southern California Coastal Water Research Project. Westminster, CA ftp://ftp.sccwrp.org/pub/download/PDFs/467_organochlorine.pdf

Switzer BC, Lewin V, Wolfe FH (1971) Shell thickness, DDE levels in eggs, and reproductive success in common terns (*Sterna hirundo*), in Alberta. Can J Zool 49:69–73

Switzer BC, Wolfe FH, Lewin V (1972) Eggshell thinning and DDE. Nature 240:162–163

Switzer BC, Lewin V, Wolfe FH (1973) DDE and reproductive success in some Alberta common terns. Can J Zool 51:1081–1086

Szaro RC (1978) Reproductive success and foraging behavior of the osprey at Seahorse Key, Florida. Wilson Bull 90:112–118

Tanabe S, Miura S, Tatsukawa R (1986) Variations of organochlorine residues with age and sex in Antarctic Minke whale. Mem Natl Inst Polar Res 44(Spec Issue):174–181

Terriere LC, Stickel LF, Keith JO, Risebrough RW, Robinson J, Gillett JW (1969) Panel discussion and open forum: impact on birds. In: Gillett JW (ed) The biological impact of pesticides in the environment. Oregon State University, Corvallis, OR, pp 65–70

Thomas S (2010) Osprey update. Tracks, a publication of the Newport Bay Naturalists & Friends, pp 1–2, March–May.

Thompson NP, Rankin PW, Cowan PE, Williams LE, Nesbitt SA (1977) Chlorinated hydrocarbon residues in the diet and eggs of the Florida brown pelican. Bull Environ Contam Toxicol 18:331–339

Tilbury KL, Stein JE, Krone CA, Brownell RL, Blokhin SA, Bolton JL, Ernest DW (2002) Chemical contaminants in juvenile gray whales (*Eschrichtius robustus*) from a subsistence harvest in Arctic feeding grounds. Chemosphere 47:239–252

Toschik PC, Rattner BA, McGowan PC, Christman MC, Carter DB, Hale RC, Matson CW, Ottinger MA (2005) Effects of contaminant exposure on reproductive success of ospreys (*Pandion haliaetus*) nesting in Delaware River and Bay, USA. Environ Toxicol Chem 24:617–628

Trimble SW (1997) Contribution of stream channel erosion to sediment yield from an urbanized watershed. Science 278:1442–1444

US Department of the Interior (1998) Guidelines for interpretation of the biological effects of selected constituents in biota, water, and sediment. National Irrigation Water Quality Program Information Report No. 3

US EPA (1980) Ambient water quality criteria for DDT, EPA 440/5-80-038
US EPA (1995) Great lakes water quality initiative criteria documents for the protection of wildlife: DDT, Mercury, 2,3,7,8-TCDD, PCBs., EPA-820-B-95-008
US EPA (1999) Fact sheet: toxaphene update: impact on fish advisories. EPA-823-F-99-018. September
US EPA Region IX (1998) Total maximum daily loads for nutrients, San Diego Creek and Newport Bay. US EPA, San Francisco, CA
US EPA Region IX (2000) Water quality standards; Establishment of numeric criteria for priority toxic pollutants for the State of California; Rule. [California Toxics Rule.] Fed Reg 65: 31682-31719
US EPA Region IX (2002) Total maximum daily loads for toxic pollutants, San Diego Creek and Newport Bay. US EPA, San Francisco, CA
US EPA Region X (2006) Columbia River Basin fish contaminant survey 1996-1998. EPA 910-R-02-006. http://yosemite.epa.gov/r10/oea.nsf/0/C3A9164ED269353788256C09005 D36B7
US FDA (2007) FDA/CFSAN fish and fisheries products hazards & controls guidance, Appendix 5, Table A-5. http://www.cfsan.fda.gov/~comm/haccp4x5.html
US GS (2010) http://www.pwrc.usgs.gov/bioeco/.htm
Varanasi U, Stein JE, Tilsbury KL, Meador JP, Sloan CA, Clark RC, Chan SL (1994) Chemical contaminants in gray whales (*Eschrichtius robustus*) stranded along the west coast of North America. Sci Total Environ 145:29–53
Vermeer K, Reynolds LM (1970) Organochlorine residues in aquatic birds in the Canadian prairie provinces. Can Field Natural 84:117–130
Vetter W, Luckas B, Heidemann G, Skirnisson K (1996) Organochlorine residues in marine mammals from the Northern Hemisphere – A consideration of the composition of organochlorine residues in the blubber of marine mammals. Sci Total Environ 186:29–39
Wafo E, Risoul V, Schembri T, Lagadec V, Dhermain F, Mama C, Portugal H (2012) PCBs and DDTs in *Stenella coeruleoalba* dolphins from the French Mediterranean coastal environment (2007-2009): current state of contamination. Mar Poll Bull 64:2535–2541
Wang D, Atkinson S, Hoover-Miller A, Lee S-E, Li QX (2007) Organochlorines in harbor seal (*Phoca vitulina*) tissues from the northern Gulf of Alaska. Environ Pollut 146:268–280
Ware GW (1975) Effects of DDT on reproduction in higher animals. Residue Rev 59:119–140
Weijs L, van Elk C, Das K, Blust R, Covaci A (2010) Persistent organic pollutants and methoxylated PBDEs in harbour porpoises from the North Sea from 1990 until 2008: young wildlife at risk? Sci Total Environ 409:228–237
Weseloh DV, Teeple SM, Gilbertson M (1983) Double-crested cormorants of the Great Lakes: egg-laying parameters, reproductive failure, and contaminant residues in eggs, Lake Huron 1972-1973. Can J Zool 61:427–436
Weseloh D, Vaughn C, Thomas W, Braune BM (1989) Organochlorine contaminants in eggs of common terns from the Canadian Great Lakes, 1981. Environ Pollut 59:141–160
White DH, Mitchell CA, Swineford DM (1984) Reproductive success of black skimmers in Texas relative to environmental pollutants. J Field Ornith 55:18–30
Wiemeyer SN, Porter RD (1970) DDE thins eggshells of captive American kestrels. Nature 227:737–738
Wiemeyer SN, Spitzer PR, Krantz WC, Lamont TG, Cromartie E (1975) Effects of environmental pollutants on Connecticut and Maryland ospreys. J Wildl Manag 39:124–139
Wiemeyer SN, Spitzer PR, McLain PD (1978) Organochlorine residues in New Jersey osprey eggs. Bull Environ Contam Toxicol 19:56–63
Wiemeyer SN, Bunck CM, Krynitsky AJ (1988) Organochlorine pesticides, polychlorinated biphenyls, and mercury in osprey eggs – 1970-79 – and their relationships to shell thinning and productivity. Arch Environ Contam Toxicol 17:767–787
Wolman AA, Wilson AJ (1970) Occurrence of pesticides in whales. Pestic Monit J 4:8–10
Woodford JE, Karasov WH, Meyer MW, Chambers L (1998) Impact of 2,3,7,8-TCDD exposure on survival, growth, and behavior of ospreys breeding in Wisconsin, USA. Environ Toxicol Chem 17:1323–1331

Word JQ, Ward JA, Franklin LM, Cullinan VI, Kiesser SL (1987) Evaluation of the equilibrium partitioning theory for estimating the toxicity of the nonpolar organic compound DDT to the sediment dwelling amphipod *Rhepozynium abronius*. Battelle Washington Environmental Program Office, Washington, DC

Wu Y, Shi J, Zheng GJ, Li P, Liang B, Chen T, Wu Y, Liu W (2013) Evaluation of organochlorine contamination in Indo-Pacific humpback dolphins (*Sousa Chinensis*) from the Pearl River Estuary, China. Sci Total Environ 444:423–429

Ylitalo GM, Stein JE, Hom T, Johnson LL, Tilbury KL, Hall AJ, Rowles T, Greig D, Lowenstine LJ, Gulland FMD (2005) The role of organochlorines in cancer-associated mortality in California sea lions (*Zalophus californianus*). Mar Pollut Bull 50:30–39

Zicus MC, Briggs MA, Pace RM (1988) DDE, PCB, and mercury residues in Minnesota common goldeneye and hooded merganser eggs, 1981. Can J Zool 66:1871–1876

INDEX FOR RECT VOL. 235

Accumulation in bivalves, paralytic shellfish toxins, **235**: 9
Accumulation rates, paralytic shellfish toxins, **235**: 10
Ag:Si (dissolved), vs. oxygen profiles in the Pacific Ocean (diags), **235**: 40
Agricultural soils in Newport Bay, DDT concentrations (table), **235**: 54
Agricultural soils, chlordane, **235**: 137
Agricultural soils, DDT, **235**: 54
Agricultural soils, toxaphene (table), **235**:146
Algal toxin, saxitoxin, **235**: 2
American avocet, DDT residues, **235**: 70
Amphipod toxicity, metals & organics in sediment (table), **235**: 142
Analysis methods, paralytic shellfish toxins, **235**: 10
Anchovies, DDE residue decreases (table), **235**: 122
Atlantic menhaden, DDT & metabolite residues (table), **235**: 123
Avian threat, DDT residues over time, **235**: 93
Benthic triad, impairment analysis, **235**: 141, 142
Bioaccumulation, DDT in Oregon, **235**: 109
Bioassay methods, paralytic shellfish toxins, **235**: 12
Biogeochemical cycling, oceanic silver, **235**: 27 ff
Biogeochemistry, silver cycling, **235**: 29
Biomagnification factor, ospreys, **235**: 114
Biotransformation, paralytic shellfish toxins, **235**: 15
Biotransformation, paralytic shellfish toxins, **235**: 1ff.
Bird eggshell thickness (table), **235**: 132

Bird residues, DDT, **235**: 61
Bivalve accumulation & depuration, paralytic shellfish toxins, **235**: 9
Bivalve detoxification, paralytic shellfish toxins (table), **235**: 11
Black skimmers, DDT residues, **235**: 69
Black-necked stilts, DDT residues, **235**: 69
Brown pelican eggs, DDE levels (diag.), **235**: 122
Brown pelican eggs, DDE residue decreases (table), **235**: 122
Brown pelican eggs, DDT & metabolites (diag.), **235**: 126
Brown pelican eggs, organochlorine residue decline, **235**: 125
Brown pelican eggshell thinning, DDE dose-response (diag.), **235**: 120
Brown pelican recovery, California (table), **235**: 121
Brown pelican reproduction, recovery (table), **235**: 124
Brown pelican reproductive effects, DDE threshold (diag.), **235**: 128
California fish guidance on DDT, human health protection, **235**: 135
California fish guidance on PCBs, human health protection, **235**:152
Canadian fish guidance, wildlife protection, **235**: 129
Cetaceans & DDT residues, P450 isozymes, **235**: 81
Chemical assay methods, paralytic shellfish toxins, **235**: 13
Chlordane hazard assessment, Newport Bay & watershed, **235**: 49 ff

Chlordane levels, water column, fish & mussels, **235**: 139
Chlordane residues, mussels (diag.), **235**: 141
Chlordane, described, **235**: 50
Chlordane, environmental levels, **235**: 137
Chlordane, *Hyallela azteca* toxicity (table), **235**: 145
Chlordane, in agricultural soils (table), **235**: 138
Chlordane, in agricultural soils, **235**: 137
Chlordane, in Newport Bay & watershed sediments (diag.), **235**: 138
Chlordane, red shiner levels (diags.), **235**: 140
Chlordane, toxicity thresholds, **235**: 144
Climate described, Newport Bay & watershed, **235**: 51
Coastal waters, dissolved silver concentrations (table), **235**: 32
Contaminant mixture issues, TEL data points, **235**: 86
Copper concentrations, Mid-Atlantic ridge plumes (table), **235**: 32
Copper, vs. silver in oceanic waters (diag.), **235**: 36
Cormorant eggs, DDE residues (table), **235**: 66
Cormorant eggshells, thickness (table), **235**: 66
Cormorants, DDT residues, **235**: 64
Cormorants, eggshell thinning & DDE (diag.), **235**: 65
DDD residues, osprey prey fish (table), **235**: 105
DDE & DDT, osprey eggshell thinning, **235**: 96
DDE bioaccumulation, Oregon, **235**: 109
DDE dose-response, brown pelican eggshell thinning (diag.), **235**: 120
DDE egg residues, osprey population status (table), **235**: 97
DDE egg residues, vs. osprey productivity (diag.), **235**: 102
DDE exposure effect, kestrel eggshell thickness (diag.), **235**: 132
DDE levels, brown pelican eggs (diag.), **235**: 122
DDE pelican egg residues, inverse eggshell thickness (diag.), **235**: 124
DDE reproductive effects, threshold for brown pelican (diag.), **235**: 128
DDE residue decline, osprey recovery, **235**: 98
DDE residue decreases, anchovies & brown pelican eggs (table), **235**: 122

DDE residue plot, brown pelican eggshell thinning (diag.), **235**: 120
DDE residues, in Texas cormorant eggs (table), **235**: 66
DDE residues, largescale suckers (diag.), **235**: 111
DDE residues, osprey egg (table), **235**: 104
DDE residues, osprey eggshell thinning (table), **235**: 110
DDE residues, osprey prey fish (table), **235**: 105
DDE residues, pelican recovery, **235**: 124
DDE residues, species variation & eggshell thinning (diag.), **235**: 119
DDE residues, vs. osprey eggshell thickness (diag.), **235**: 102
DDE residues, vs. pelican eggshell thickness (diag.), **235**: 127
DDE, eggshell thinning of pelicans, **235**: 116
DDE, pelican eggshell thinning (diag.), **235**: 118
DDT & eggshell thickness, birds (table), **235**: 132
DDT & human health, California fish guidance, **235**: 135
DDT & metabolites, brown pelican eggs (diag.), **235**: 126
DDT & metabolites, in emitted Atlantic menhaden (table), **235**: 123
DDT & peregrine falcons, wintering ground exposures (table), **235**: 106
DDT & porpoises, stress hormone induction, **235**: 79
DDT agricultural soil samples, locations (illus.), **235**: 56
DDT concentration trends, fish species, **235**: 59
DDT concentrations, agricultural soils, **235**: 54
DDT concentrations, Newport Bay agricultural soils (table), **235**: 54
DDT dissipation, process summarized, **235**: 54
DDT effect data sets, fresh & marine water sediments (table), **235**: 85
DDT effect discrepancies, fresh- & marine-water sediments, **235**: 85
DDT hazard assessment, Newport Bay & watershed, **235**: 49 ff.
DDT levels, in the environment, **235**: 53
DDT levels, Newport Bay & watershed water column (table), **235**: 58
DDT release proximity, dolphin residues, **235**: 79

DDT residue date, in sediments, **235**: 55
DDT residue differences, stranded vs. harvested whales, **235**: 80
DDT residue trends, fish, **235**: 103
DDT residues in cetaceans, P450 isozymes, **235**: 81
DDT residues in whales, gender differences, **235**: 81
DDT residues over time, avian threat, **235**: 93
DDT residues, American avocet, **235**: 70
DDT residues, and soil erosion, **235**: 55
DDT residues, black skimmers, **235**: 69
DDT residues, black-necked stilts, **235**: 69
DDT residues, cormorants, **235**: 64
DDT residues, dolphins, **235**: 78
DDT residues, in birds, **235**: 61
DDT residues, in fish (table), **235**: 94
DDT residues, in fish of Great Gull island (table), **235**: 91
DDT residues, in marine mammals, **235**: 82
DDT residues, in Montana osprey eggs (table), **235**: 100
DDT residues, in Newport Bay & watershed mussels (diag.), **235**: 60
DDT residues, in osprey eggs (table), **235**: 93
DDT residues, in sea lions, **235**: 75
DDT residues, in seals, **235**: 73
DDT residues, in whales, **235**: 80
DDT residues, killdeer, **235**: 70
DDT residues, Newport Bay region dolphins, **235**: 80
DDT residues, Newport Bay sediment data (diag.), **235**: 57
DDT residues, osprey egg (table), **235**: 104
DDT residues, ospreys, **235**: 93
DDT residues, terns, **235**: 62
DDT residues, vs. toxicity in marine sediments, **235**: 87
DDT residues, water column, fish & mussels, **235**: 58
DDT residues, white sucker tissues (diag.), **235**: 103
DDT safe levels, fresh water organisms, **235**: 88
DDT safety, NAS recommendation, **235**: 92
DDT TEL calculation, issues, **235**: 89
DDT TEL levels, in sediment, **235**: 83
DDT TEL values, in Newport Bay region, **235**: 84
DDT TMDL targets, in sediment, fish & water (table), **235**: 83
DDT TMDL values, in Newport Bay & watershed, **235**: 83

DDT toxicity, vs. residues, **235**: 87
DDT, and eggshell thinning, **235**: 61
DDT, described, **235**: 50
DDT, in marine mammals, **235**: 72
DDT, pelicans, **235**: 115
DDT, red shiner residues in San Diego Creek (diag.), **235**: 58
DDT, wildlife protection guidance, **235**: 114
DDT:DDE egg residue ratio, exposure timing, **235**: 107
Depuration in bivalves, paralytic shellfish toxins, **235**: 9
Detoxification in bivalves, paralytic shellfish toxins (table), **235**: 11
Detoxification rates, paralytic shellfish toxins, **235**: 10
Detoxification, paralytic shellfish toxins, **235**:
Dissolved silver concentrations, estuarine & coastal waters (table), **235**: 32
Dolphin residues, of DDT, **235**: 78
Dolphin residues, proximity to DDT release, **235**: 79
Dolphins & DDT residues, from Newport Bay, **235**: 80
Effects range median (ERM), chlordane (table), **235**: 143
Eggshell thickness & DDT, birds (table), **235**: 132
Eggshell thickness, changes in New Jersey (table), **235**: 97
Eggshell thickness, osprey eggs (table), **235**: 104
Eggshell thickness, vs. DDE residues, **235**: 124
Eggshell thinning & DDE, cormorants (diag.), **235**: 65
Eggshell thinning & species variation, DDE residues (diag.), **235**: 119
Eggshell thinning in brown pelican, DDE dose-response (diag.), **235**: 120
Eggshell thinning in brown pelicans, DDE residue plot (diag.), **235**: 120
Eggshell thinning in ospreys, DDE residues (table), **235**: 110
Eggshell thinning in ospreys, DDE, **235**: 96
Eggshell thinning of pelicans, DDE, **235**: 116
Eggshell thinning, and DDT, **235**: 61
Environmental effects, harmful algal blooms, **235**: 7
Environmental levels, chlordane, **235**: 137
Environmental levels, toxaphene, **235**:146
Environmental release, silver, **235**: 28

ERM (effects range median), chlordane (table), **235**: 143
Estuarine waters, dissolved silver concentrations (table), **235**: 32
Eutrophication, influences on harmful algal blooms, **235**: 6
Exposure timing, DDT:DDE egg residue levels, **235**: 107
Fish concentration trends, DDT, **235**: 59
Fish DDT residue trends, 1970's, **235**: 103
Fish fillets from Newport Bay, PCB residues (table), **235**:151
Fish levels, chlordane, **235**: 139
Fish residues, DDT (table), **235**: 94
Fish residues, DDT in Newport Bay area, **235**: 58
Fish residues, of DDT (table), **235**: 91
Fish residues, toxaphene, **235**:147
Fish TMDL targets, for DDT (table), **235**: 83
Fluxes, of oceanic silver, **235**: 27 ff.
Fresh water safe levels, DDT residues, **235**: 88
Fresh water sediments, effect data sets for total DDT (table), **235**: 85
Great Lakes, osprey recovery (diags.), **235**: 108
Harmful algal bloom species, toxins produced (table), **235**: 5
Harmful algal blooms, defined, **235**: 2
Harmful algal blooms, environmental effects, **235**: 7
Harmful algal blooms, eutrophication effects, **235**: 6
Harmful algal blooms, factors influencing toxicity, **235**: 6
Harmful algal blooms, light intensity effects, **235**: 8
Harmful algal blooms, nutrient effects, **235**: 7
Harmful algal blooms, temperature effects, **235**: 8
Human hazard assessment, DDT in Newport Bay & watershed, **235**: 49 ff.
Human health protection & DDT, California fish guidance, **235**: 135
Human health protection, California guidance on PCBs, **235**:152
Hyallela azteca toxicity, chlordane (table), **235**: 145
Hydrology described, of Newport Bay and watershed, **235**: 51
Immunosorbant assays, paralytic shellfish toxins, **235**: 14
Intoxication symptoms, paralytic shellfish toxins, **235**: 2
Kestrel eggshell thickness, DDE exposure effect (diag.), **235**: 132

Killdeer, DDT residues, **235**: 70
Light intensity effects, harmful algal blooms, **235**: 8
Marine mammals, and DDT residues, **235**: 82
Marine mammals, DDT residues, **235**: 72
Marine sediment, silver cycling, **235**: 39
Marine sediments, silver concentrations (table), **235**: 31
Marine toxins, paralytic shellfish, **235**: 1ff.
Marine water sediments, effect data sets for total DDT (table), **235**: 85
Metals in sediments, amphipod toxicity (table), **235**: 142
Mid-Atlantic ridge plumes, Ag, Cu & silicate concentrations (table), **235**: 32
Mode of action, paralytic shellfish toxins, **235**: 2
Mussel levels in Newport Bay region, toxaphene (diag.), **235**:148
Mussel levels, chlordane (diag.), **235**: 141
Mussel levels, chlordane, **235**: 139
Mussel residues of DDT, Newport Bay & watershed (diag.), **235**: 60
Mussel residues, DDT in Newport Bay area, **235**: 58
Mussels, toxaphene, **235**:147
NAS (National Academy of Science) guidance, to protect wildlife, **235**: 90
NAS fish guidance on toxaphene, wildlife protection, **235**:149
NAS recommendation, DDT safety, **235**: 92
National Academy of Science (NAS), wildlife protection, **235**: 90
New York sediment guidance, toxaphene, **235**:149
Newport Bay & watershed mussels, DDT residues (diag.), **235**: 60
Newport Bay & watershed water column, DDT levels (table), **235**: 58
Newport Bay & watershed, chlordane in mussels (diag.), **235**: 141
Newport Bay & watershed, DDT, chlordane, toxaphene & PCB hazard assessment, **235**: 49 ff.
Newport Bay & watershed, map (illus.), **235**: 51
Newport Bay & watershed, TMDL values for DDT, **235**: 83
Newport Bay agricultural soils, DDT concentrations (table), **235**: 54
Newport Bay and watershed, described, **235**: 51
Newport Bay area, DDT in fish & mussels, **235**: 58

Newport Bay area, DDT residues in birds, **235**: 61
Newport Bay dolphins, DDT residues, **235**: 80
Newport Bay region sediment, chlordane levels (diag.), **235**: 138
Newport Bay region, DDT TEL values, **235**: 84
Newport Bay sediment date, DDT residues (diag.), **235**: 57
Newport Bay, land use, **235**: 52
Newport Bay, water quality, **235**: 52
Nutrient effects, harmful algal blooms, **235**: 7
Oceanic contaminant, silver, **235**: 28
Oceanic contamination profile, silver (illus.), **235**: 31
Oceanic contamination, by silver (table), **235**: 30
Oceanic contour plot, silver concentrations (illus.), **235**: 37
Oceanic data, silver contamination, **235**: 29
Oceanic fixation, silver, **235**: 41
Oceanic silver, data comparisons, **235**: 42
Oceanic silver, measurements, **235**: 33
Oceanic silver, vertical concentration profiles (diag.), **235**: 34
Organ elimination, bivalve paralytic shellfish toxins, **235**: 10
Organics in sediments, amphipod toxicity (table), **235**: 142
Organochlorine residue decline, brown pelican eggs, **235**: 125
Osprey egg residues, DDE & DDD (table), **235**: 104
Osprey eggs, DDT residues (table), **235**: 100
Osprey eggs, DDT residues (table), **235**: 93
Osprey eggs, nesting productivity (table), **235**: 100
Osprey eggshell thickness, vs. DDE residues (diag.), **235**: 102
Osprey eggshell thinning, DDE residues (table), **235**: 110
Osprey population status, DDE egg residues (table), **235**: 97
Osprey populations, reproductive success (table), **235**: 95
Osprey prey fish, DDE & DDD residues (table), **235**: 105
Osprey productivity, New Jersey (table), **235**: 105
Osprey productivity, vs. DDE egg residues (diag.), **235**: 102
Osprey recovery, from DDT effects, **235**: 98
Osprey recovery, Great Lakes (diags.), **235**: 108

Osprey recovery, in Connecticut-Long Island (diag.), **235**: 99
Osprey reproduction, Newport Bay, **235**: 112
Osprey reproduction, San Joaquin Wildlife Sanctuary (table), **235**: 113
Ospreys & DDE, eggshell thinning, **235**: 96
Ospreys, biomagnifications factor, **235**: 114
Ospreys, DDT residues, **235**: 93
Ospreys, declining populations, **235**: 94
Ospreys, eggshell thickness (table), **235**: 104
Ospreys, population size & productivity (diag.), **235**: 101
Ospreys, population status (table), **235**: 95
Ospreys, wintering locations (illus.), **235**: 111
P450 isozymes, cetaceans & DDT residues, **235**: 81
Pacific Ocean oxygen profiles, vs. Ag:Si (dissolved) (diags.), **235**: 40
Paralytic shell fish toxins, organisms responsible, **235**: 3
Paralytic shellfish toxin composition, salinity effect, **235**: 8
Paralytic shellfish toxins, analysis methods, **235**: 10
Paralytic shellfish toxins, bioassay methods, **235**: 12
Paralytic shellfish toxins, biotransformation & detoxification, **235**: 1ff.
Paralytic shellfish toxins, biotransformation, **235**: 15
Paralytic shellfish toxins, chemical assay methods, **235**: 13
Paralytic shellfish toxins, clinical symptoms, **235**: 2
Paralytic shellfish toxins, detoxification in bivalves (table), **235**: 11
Paralytic shellfish toxins, immunosorbant assays, **235**: 14
Paralytic shellfish toxins, mode of action, **235**: 2
Paralytic shellfish toxins, organ elimination & accumulation rates, paralytic shellfish toxins, **235**: 10
Paralytic shellfish toxins, paralytic shellfish poisoning, **235**: 2
Paralytic shellfish toxins, structure & toxicity (diag., table), **235**: 4
Paralytic shellfish toxins, treatments, **235**: 3
PCB (polychlorinated biphenyls), sport fish levels, **235**:150
PCB hazard assessment, Newport Bay & watershed, **235**: 49 ff
PCB residues, fish fillets from Newport Bay (table), **235**:151

PCBs, California guidance to protect human health, **235**:152
PCBs, described, **235**: 50
Pelican eggshell thickness, vs. DDE residues (diag.), **235**: 127
Pelican eggshell thickness, vs. DDE residues, **235**: 124
Pelican eggshell thinning, DDE (diag.), **235**: 118
Pelican eggshell thinning, DDE, **235**: 116
Pelican recovery, DDE residues, **235**: 124
Pelicans, DDT, **235**: 115
Peregrine falcons & DDT, wintering ground exposures (table), **235**: 106
Polychlorinated biphenyls (PCBs), sport fish levels, **235**:150
Population status, ospreys (table), **235**: 95
Porpoises & DDT, stress hormone induction, **235**: 79
Red shiner levels, chlordane (diags.), **235**: 140
Red shiner levels, toxaphene (diags.), **235**:147, 148
Red shiner residues in San Diego Creek, DDT (diag.), **235**: 59
Red tide, described, **235**: 1
Reproductive success, U.S. osprey populations (table), **235**: 95
Salinity effect, paralytic shellfish toxin composition, **235**: 8
Salinity effect, silver concentration in estuarine waters, **235**: 33
San Joaquin Wildlife Sanctuary, osprey reproduction (table), **235**: 113
Saxitoxin, algal toxin, **235**: 2
Saxitoxins, categories, **235**: 3
Sea lions, DDT residues, **235**: 75
Seals, DDT residues, **235**: 73
Sediment data, DDT in Newport Bay (diag.), **235**: 57
Sediment date, DDT, **235**: 55
Sediment guidance, toxaphene, **235**:149
Sediment levels, chlordane, **235**: 138
Sediment loads, vs. fate of organochlorines, **235**: 53
Sediment TMDL targets, for DDT (table), **235**: 83
Sediments, toxaphene levels, **235**:146
Selenium concentrations, vs. silver in water & aerosols (diags.), **235**: 38
Silicate concentrations, Mid-Atlantic ridge plumes (table), **235**: 32
Silicate, vs. silver in oceanic waters (diag.), **235**: 35

Silver concentrations in N. Pacific, temporal variations (diag.), **235**: 37
Silver concentrations, marine sediments (table), **235**: 31
Silver concentrations, Mid-Atlantic ridge plumes (table), **235**: 32
Silver concentrations, oceanic contour plot (illus.), **235**: 37
Silver concentrations, seawater reference material (table), **235**: 43
Silver concentrations, ultraviolet light effects, **235**: 42
Silver concentrations, vs. selenium in water & aerosols (diags.), **235**: 38
Silver contaminant profiles, oceans (illus.), **235**: 31
Silver contamination, oceanic data, **235**: 29
Silver contamination, oceans (table), **235**: 30
Silver cycling, marine sediment, **235**: 39
Silver fixation, ocean, **235**: 41
Silver in the oceans, data comparisons, **235**: 42
Silver oceanic distribution, low oxygen effect, **235**: 39
Silver profiles, vertical oceanic concentrations (diag.), **235**: 34
Silver, as oceanic contaminant, **235**: 28
Silver, biogeochemical cycling, **235**: 29
Silver, early production and use, **235**: 28
Silver, oceanic biogeochemical cycling, **235**: 27 ff.
Silver, oceanic sources & fluxes, **235**: 27 ff.
Silver, sources in environment, **235**: 28
Silver, toxicity issues, **235**: 28
Silver, vs. dissolved copper in oceanic waters (diag.), **235**: 36
Silver, vs. silicate in oceanic waters (diag.), **235**: 35
Silver, water concentrations, **235**: 33
Soil erosion, and DDT residues, **235**: 55
Sources & cycling, oceanic silver, **235**: 27 ff.
Species variation & eggshell thinning, DDE residues (diag.), **235**: 119
Stress hormone induction, DDT in porpoises, **235**: 79
Suckers (largescale), & DDE residues (diag.), **235**: 111
TEL (threshold effect levels) for DDT, in sediment, **235**: 83
TEL calculation issues, DDT residues, **235**: 89
TEL data points, contaminant mixtures, **235**: 86
TEL use issues, TMDL sediment target, **235**: 90

TEL values for DDT, in Newport Bay region, **235**: 84
Temperature effects, harmful algal blooms, **235**: 8
Temporal variations, silver concentrations in N. Pacific (diag.), **235**: 37
Terns, DDT residues, **235**: 62
Threshold effect levels (TELs) for DDT, in sediment, **235**: 83
TMDL (total maximum daily load) targets for DDT, in Newport Bay & watershed, **235**: 83
TMDL targets for DDT, in sediment, fish & water (table), **235**: 83
Total maximum daily load (TMDL) targets for DDT, in Newport Bay & watershed, **235**: 83
Toxaphene decay, influencing factors, **235**:149
Toxaphene levels, Newport Bay & watershed mussels (table), **235**:148
Toxaphene levels, sediments, **235**:146
Toxaphene residues, water column, fish & mussels, **235**:147
Toxaphene, described, **235**: 50
Toxaphene, environmental levels, **235**:146
Toxaphene, hazard assessment, Newport Bay & watershed, **235**: 49 ff
Toxaphene, in agricultural soils (table), **235**:146
Toxaphene, NAS fish guidance, **235**:149
Toxaphene, New York sediment guidance, **235**:149
Toxaphene, red shiner levels (diags), **235**:147, 148
Toxicity issues, silver, **235**: 28
Toxicity of harmful algal blooms, influencing factors, **235**: 6
Toxicity thresholds, chlordane, **235**: 144
Toxicity, paralytic shellfish toxins (table), **235**:4
Treatment, paralytic shellfish toxins, **235**: 3
U.S. EPA fish guidance, wildlife protection, **235**: 134
Ultraviolet light effects, silver concentrations, **235**: 42
Water column guidance on DDT, to protect wildlife, **235**: 114
Water column levels, chlordane, **235**: 139
Water column of Newport Bay & watershed, DDT levels (table), **235**: 58
Water column, toxaphene, **235**:147
Water concentrations, silver, **235**: 33
Water quality, Newport Bay, **235**: 52
Water TMDL targets, for DDT (table), **235**: 83
Whale DDT residues, stranded vs. harvested specimens, **235**: 80
Whales, DDT residues, **235**: 80
White sucker tissues, DDT residues (diag.), **235**: 103
Wildlife hazard assessment, DDT in Newport Bay & watershed, **235**: 49 ff.
Wildlife protection, Canadian fish guidance, **235**: 129
Wildlife protection, NAS guidance, **235**: 90
Wildlife protection, U.S. EPA fish guidance, **235**: 134
Wildlife protection, U.S. EPA water column guidance on DDT, **235**: 114
Wintering ground exposures to DDT, peregrine falcons (table), **235**: 106